Basics of
NANOTECHNOLOGY

Basics of
NANOTECHNOLOGY

Sidharth Baliyan

ANMOL PUBLICATIONS PVT. LTD.
NEW DELHI-110 002 (INDIA)

ANMOL PUBLICATIONS PVT. LTD.
Regd. Office: 4360/4, Ansari Road, Daryaganj,
New Delhi-110002 (India)
Tel.: 23278000, 23261597, 23286875, 23255577
Fax: 91-11-23280289
Email: anmolpub@gmail.com
Visit us at: www.anmolpublications.com

Branch Office: No. 1015, Ist Main Road, BSK IIIrd Stage
IIIrd Phase, IIIrd Block, Bengaluru-560 085 (India)
Tel.: 080-41723429 • Fax: 080-26723604
Email: anmolpublicationsbangalore@gmail.com

Basics of Nanotechnology

First Edition, 2011

ISBN 978-81-261-4904-9

PRINTED IN INDIA

Printed at Suman Printers, Delhi

Contents

Preface

Nanotechnology is the engineering of tiny machines - the projected ability to build things from the bottom up inside personal nanofactories (PNs), using techniques and tools being developed today to make complete, highly advanced products. Ultimately, nanotechnology will enable control of matter at the nanometer scale, using mechanochemistry. Shortly after this envisioned molecular machinery is created, it will result in a manufacturing revolution, probably causing severe disruption. It also has serious economic, social, environmental, and military implications.

It's worth pointing out that the word "nanotechnology" has become very popular and is used to describe many types of research where the characteristic dimensions are less than about 1,000 nanometers. For example, continued improvements in lithography have resulted in line widths that are less than one micron: this work is often called "nanotechnology." Sub-micron lithography is clearly very valuable (ask anyone who uses a computer!) but it is equally clear that conventional lithography will not let us build semiconductor devices in which individual dopant atoms are located at specific lattice sites. Many of the exponentially improving trends in computer hardware capability have remained steady for the last 50 years. There is fairly widespread belief that these trends are likely to continue for at least another several years, but then conventional lithography starts to reach its limits.

Author

1

Nanometers

INTRODUCTION

A nanometer is one billionth of a meter (10–9 m) and is the unit of length that is generally most appropriate for describing the size of single molecules. The nanoscale marks the nebulous boundary between the classical and quantum mechanical worlds; thus, realization of nanotechnology promises to bring revolutionary capabilities. Fabrication of nanomachines, nanoelectronics and other nanodevices will undoubtedly solve an enormous amount of the problems faced by mankind today. Nanotechnology is currently in a very infantile stage. However, we now have the ability to organize matter on the atomic scale and there are already numerous products available as a direct result of our rapidly increasing ability to fabricate and characterize feature sizes less than 100 nm.

Mirrors that don't fog, biomimetic paint with a contact angle near 180°, gene chips and fat soluble vitamins in aqueous beverages are some of the first manifestations of nanotechnology. However, immenant breakthroughs in computer science and medicine will be where the real potential of nanotechnology will first be achieved. Nanoscience is an interdisciplinary field that seeks to bring about mature nanotechnology.

Focusing on the nanoscale intersection of fields such as physics, biology, engineering, chemistry, computer science and more, nanoscience is rapidly expanding. Nanotechnology centers are popping up around the world as more funding is provided and nanotechnology market share increases.

Nanotechnology is extremely diverse, ranging from novel extensions of conventional device physics, to completely new approaches based upon molecular self-assembly, to developing new materials with dimensions on the nanoscale, even to speculation on whether we can directly control matter on the atomic scale. There has been much debate on the future of implications of nanotechnology.

Nanotechnology has the potential to create many new materials and devices with wide-ranging applications, such as in medicine, electronics, and energy production. On the other hand, nanotechnology raises many of the same issues as with any introduction of new technology, including concerns about the toxicity and environmental impact of nanomaterials, and their potential effects on global economics, as well as speculation about various doomsday scenarios.

These concerns have led to a debate among advocacy groups and governments on whether special regulation of nanotechnology is warranted. Feynman described a process by which the ability to manipulate individual atoms and molecules might be developed, using one set of precise tools to build and operate another proportionally smaller set, so on down to the needed scale.

CONCEPTS

One nanometer (nm) is one billionth, or 10^{-9}, of a meter. By comparison, typical carbon-carbon bond lengths, or the spacing between these atoms in a molecule, are in the range 0.12-0.15 nm, and a DNA double-helix has a diameter around 2 nm. On the other hand, the smallest cellular lifeforms, the bacteria of the genus Mycoplasma, are around 200 nm in length. To put that scale in another context, the comparative size of a nanometer to a meter is the same as that of a marble to the size of the earth. Or another way of putting it: a nanometer is the amount a man's beard grows in the time it takes him to raise the razor to his face.

Two main approaches are used in nanotechnology. In the "bottom-up" approach, materials and devices are built from molecular components which assemble themselves chemically by principles of molecular recognition. In the "top-down" approach, nano-objects are constructed from larger entities without atomic-level control.

COMPARISION OF MATERIALS

Additionally, a number of physical (mechanical, electrical, optical, etc.) properties change when compared to macroscopic systems. One example is the increase in surface area to volume ratio altering mechanical, thermal and catalytic properties of materials. Novel *mechanical* properties of nanosystems are of interest in the nanomechanics research. The catalytic activity of nanomaterials also opens potential risks in their interaction with biomaterials. Materials reduced to the nanoscale can show different properties compared to what they exhibitf on a macroscale, enabling unique applications. For instance, opaque substances become transparent (copper); stable materials turn combustible (aluminum); solids turn into liquids at room temperature (gold); insulators become conductors (silicon).

A material such as gold, which is chemically inert at normal scales, can serve as a potent chemical catalyst at nanoscales. Much of the fascination with nanotechnology stems from these quantum and surface phenomena that matter exhibits at the nanoscale. Modern synthetic chemistry has reached the point where it is possible to prepare small molecules to almost any structure. These methods are used today to produce a wide variety of useful chemicals such as pharmaceuticals or commercial polymers.

This ability raises the question of extending this kind of control to the next-larger level, seeking methods to assemble these single molecules into supramolecular assemblies consisting of many molecules arranged in a well defined manner. These approaches utilize the concepts of molecular self-assembly and/or supramolecular chemistry to automatically arrange themselves into some useful conformation through a bottom-up approach. The concept of molecular recognition is especially important: molecules can be designed so that a specific conformation or arrangement is favoured due to non-covalent intermolecular forces.

The Watson-Crick basepairing rules are a direct result of this, as is the specificity of an enzyme being targeted to a single substrate, or the specific folding of the protein itself. Thus, two or more components can be designed to be complementary and mutually attractive so that they make a more complex and useful whole. Such bottom-up approaches should be able to produce devices in parallel and much cheaper than top-down methods, but could potentially be overwhelmed as the size and complexity of the desired assembly increases. Most useful structures require complex and thermody-namically unlikely arrangements of atoms. Nevertheless, there are many examples of self-assembly based on molecular recognition in biology, most notably Watson-Crick basepairing and enzyme-substrate interactions. The challenge for nanotechnology is whether these principles can be used to engineer novel constructs in addition to natural ones.

MOLECULAR NANOTECHNOLOGY

Molecular nanotechnology, sometimes called molecular manufacturing, is a term given to the concept of engineered nanosystems (nanoscale machines) operating on the molecular scale. It is especially associated with the concept of a molecular assembler, a machine that can produce a desired structure or device atom-by-atom using the principles of mechanosynthesis. Manufacturing in the context of productive nanosystems is not related to, and should be clearly distinguished from, the conventional technologies used to manufacture nanomaterials such as carbon nanotubes and nanoparticles. When the term "nanotechnology" was independently coined and popularized by Eric Drexler it referred to a future manufacturing technology based on molecular machine systems.

The premise was that molecular scale biological analogies of traditional machine components demonstrated molecular machines were possible: by the countless examples found in biology, it is known that sophisticated, stochastically optimised biological machines can be produced. It is hoped that developments in nanotechnology will make possible their construction by some other means, perhaps using biomimetic principles. However, Drexler and other researchers have proposed that advanced nanotechnology, although perhaps initially implemented by biomimetic means, ultimately could be based on mechanical engineering principles, namely, a manufacturing technology based on the mechanical functionality of these components (such as gears, bearings, motors, and structural members) that would enable programmable, positional assembly to atomic specification. The physics and engineering performance of exemplar designs were analysed in Drexler's book *Nanosystems*. In general it is very difficult to assemble devices on the atomic scale, as all one has to position atoms are other atoms of comparable size and stickyness.

TOOLS

A microfabricated cantilever with a sharp tip is deflected by features on a sample surface, much like in a phonograph but on a much smaller scale. A laser beam reflects off the backside of the cantilever into a set of photodetectors, allowing the deflection to be measured and assembled into an image of the surface.The first observations and size measurements of nano-particles were made during the first decade of the 20th century. They are mostly associated with the name of Zsigmondy who made detailed studies of gold sols and other nanomaterials with sizes down to 10 nm and less. He published a book in 1914. He used ultramicroscope that employs a *dark field* method for seeing particles with sizes much less than light wavelength.

There are traditional techniques developed during 20th century in Interface and Colloid Science for characterizing nanomaterials. These are widely used for *first generation* passive nanomaterials specified in the next section. These methods include several different techniques for characterizing particle size distribution. This characterization is imperative because many materials that are expected to be nano-sized are actually aggregated in solutions. Some of methods are based on light scattering. Other apply ultrasound, such as ultrasound attenuation spectroscopy for testing concentrated nanodispersions and microemulsions. There is also a group of traditional techniques for characterizing surface charge or zeta potential of nano-particles in solutions.

This information is required for proper system stabilzation, preventing its aggregation or flocculation. These methods include microelectrophoresis, electrophoretic light scattering and electroacoustics. The last one, for instance colloid vibration current method is suitable for characterizing concentrated systems. Next group of nanotechnological techniques include those used for fabrication of nanowires, those used in semiconductor fabrication such as deep ultraviolet lithography, electron beam lithography, focused ion beam machining, nanoimprint lithography, atomic layer deposition, and molecular vapour deposition, and further including molecular self-assembly techniques such as those employing di-block copolymers.

However, all of these techniques preceded the nanotech era, and are extensions in the development of scientific advancements rather than techniques which were devised with the sole purpose of creating nanotechnology and which were results of nanotechnology research. There are several important modern developments. The atomic force microscope (AFM) and the Scanning Tunneling Microscope (STM) are two early versions of scanning probes that launched nanotechnology.

There are other types of scanning probe microscopy, all flowing from the ideas of the scanning confocal microscope developed by Marvin Minsky in 1961 and the scanning acoustic microscope (SAM) developed by Calvin Quate and coworkers in the 1970s, that made it possible to see structures at the nanoscale. The tip of a scanning probe can also be used to manipulate nanostructures (a process called positional assembly). Feature-oriented scanning-positioning methodology suggested by Rostislav Lapshin appears to be a promising way to implement these nanomanipulations in automatic mode.

However, this is still a slow process because of low scanning velocity of the microscope. Various techniques of nanolithography such as dip pen nanolithography, electron beam lithography or nanoimprint lithography were also developed. Lithography is a top-down fabrication technique where a bulk material is reduced in size to nanoscale pattern. The top-down approach anticipates nanodevices that must be built piece by piece in stages, much as manufactured items are made. Scanning probe microscopy is an important technique both for characterization and synthesis of nanomaterials. Atomic force microscopes and scanning tunneling microscopes can be used to look at surfaces and to move atoms around.

By designing different tips for these microscopes, they can be used for carving out structures on surfaces and to help guide self-assembling structures. By using, for example, feature-oriented scanning-positioning approach, atoms can be moved around on a surface with scanning probe microscopy techniques.

At present, it is expensive and time-consuming for mass production but very suitable for laboratory experimentation. In contrast, bottom-up techniques build or grow larger structures atom by atom or molecule by molecule. These techniques include chemical synthesis, self-assembly and positional assembly. Another variation of the bottom-up approach is molecular beam epitaxy or MBE.

Researchers at Bell Telephone Laboratories like John R. Arthur. Alfred Y. Cho, and Art C. Gossard developed and implemented MBE as a research tool in the late 1960s and 1970s. Samples made by MBE were key to the discovery of the fractional quantum Hall effect for which the 1998 Nobel Prize in Physics was awarded. MBE allows scientists to lay down atomically-precise layers of atoms and, in the process, build up complex structures. Important for research on semiconductors, MBE is also widely used to make samples and devices for the newly emerging field of spintronics.

Newer techniques such as Dual Polarisation Interferometry are enabling scientists to measure quantitatively the molecular interactions that take place at the nano-scale. However, new therapeutic products, based on responsive nanomaterials, such as the ultradeformable, stress-sensitive Transfersome vesicles, are under development and already approved for human use in some countries.

APPLICATIONS

Their optical properties, *e.g.* fluorescence, become a function of the particle diameter. When brought into a bulk material, nanoparticles can strongly influence the mechanical properties of the material, like stiffness or elasticity. For example, traditional polymers can be reinforced by nanoparticles resulting in novel materials which can be used as lightweight replacements for metals. Therefore, an increasing societal benefit of such nanoparticles can be expected. Such nanotechnologically enhanced materials will enable a weight reduction accompanied by an increase in stability and an improved functionality. There are many applications of nano technology, few of them are shown here.

MEDICINE

The biological and medical research communities have exploited the unique properties of nanomaterials for various applications (*e.g.*, contrast agents for cell imaging and therapeutics for treating cancer). Terms such as *biomedical nanotechnology, bionanotechnology,* and *nanomedicine* are used to describe this hybrid field. Functionalities can be added to nanomaterials by interfacing them with biological molecules or structures. The size of nanomaterials is similar to that of most biological molecules and structures; therefore, nanomaterials can be useful for both in vivo and in vitro biomedical research and applications. Thus far, the integration of nanomaterials with biology has led to the development of diagnostic devices, contrast agents, analytical tools, physical therapy applications, and drug delivery vehicles.

DIAGNOSTICS

Nanotechnology-on-a-chip is one more dimension of lab-on-a-chip technology. Biological tests measuring the presence or activity of selected substances become quicker, more sensitive and more flexible when certain nanoscale particles are put to work as tags or labels.

Magnetic nanoparticles, bound to a suitable antibody, are used to label specific molecules, structures or microorganisms. Gold nanoparticles tagged with short segments of DNA can be used for detection of genetic sequence in a sample. Multicolor optical coding for biological assays has been achieved by embedding different-sized quantum dots into polymeric microbeads. Nanopore technology for analysis of nucleic acids converts strings of nucleotides directly into electronic signatures.

DRUG DELIVERY

The overall drug consumption and side-effects can be lowered significantly by depositing the active agent in the morbid region only and in no higher dose than needed. This highly selective approach reduces costs and human suffering. An example can be found in dendrimers and nanoporous materials.

They could hold small drug molecules transporting them to the desired location. Another vision is based on small electromechanical systems; NEMS are being investigated for the active release of drugs. Some potentially important applications include cancer treatment with iron nanoparticles or gold shells. A targeted or personalized medicine reduces the drug consumption and treatment expenses resulting in an overall societal benefit by reducing the costs to the public health system.

Nanotechnology is also opening up new opportunities in implantable delivery systems, which are often preferable to the use of injectable drugs, because the latter frequently display first-order kinetics (the blood concentration goes up rapidly, but drops exponentially over time). This rapid rise may cause difficulties with toxicity, and drug efficacy can diminish as the drug concentration falls below the targeted range.

TISSUE ENGINEERING

Nanotechnology can help to reproduce or to repair

damaged tissue. This so called "tissue engineering" makes use of artificially stimulated cell proliferation by using suitable nanomaterial-based scaffolds and growth factors. Tissue engineering might replace today's conventional treatments like organ transplants or artificial implants. On the other hand, tissue engineering is closely related to the ethical debate on human stem cells and its ethical implications.

CHEMISTRY AND ENVIRONMENT

Chemical catalysis and filtration techniques are two prominent examples where nanotechnology already plays a role. The synthesis provides novel materials with tailored features and chemical properties: for example, nanoparticles with a distinct chemical surrounding (ligands), or specific optical properties. In this sense, chemistry is indeed a basic nanoscience. In a short-term perspective, chemistry will provide novel "nanomaterials" and in the long run, superior processes such as "self-assembly" will enable energy and time preserving strategies.

In a sense, all chemical synthesis can be understood in terms of nanotechnology, because of its ability to manufacture certain molecules. Thus, chemistry forms a base for nanotechnology providing tailor-made molecules, polymers, etcetera, as well as clusters and nanoparticles. Chemical catalysis benefits especially from nanoparticles, due to the extremely large surface to volume ratio. The application potential of nanoparticles in catalysis ranges from fuel cell to catalytic converters and photocatalytic devices. Catalysis is also important for the production of chemicals.

Platinum nanoparticles are now being considered in the next generation of automotive catalytic converters because the very high surface area of nanoparticles could reduce the amount of platinum required. However, some concerns have been raised due to experiments demonstrating that they will spontaneously combust if methane is mixed with the ambient air. Ongoing research at the Centre National de la Recherche Scientifique (CNRS) in France may resolve their true usefulness for catalytic applications. Nanofiltration may come to be an important application, although future research must be careful to investigate possible toxicity.

A strong influence of nanochemistry on waste-water treatment, air purification and energy storage devices is to be expected. Mechanical or chemical methods can be used for effective filtration techniques.

One class of filtration techniques is based on the use of membranes with suitable hole sizes, whereby the liquid is pressed through the membrane. Nanoporous membranes are suitable for a mechanical filtration with extremely small pores smaller than 10 nm ("nanofiltration") and may be composed of nanotubes. Nanofiltration is mainly used for the removal of ions or the separation of different fluids. On a larger scale, the membrane filtration technique is named ultrafiltration, which works down to between 10 and 100 nm.

One important field of application for ultrafiltration is medical purposes as can be found in renal dialysis. Magnetic nanoparticles offer an effective and reliable method to remove heavy metal contaminants from waste water by making use of magnetic separation techniques. Using nanoscale particles increases the efficiency to absorb the contaminants and is comparatively inexpensive compared to traditional precipitation and filtration methods. Some water-treatment devices incorporating nanotechnology are already on the market, with more in development. Low-cost nanostructured separation membranes methods have been shown to be effective in producing potable water in a recent study.

ENERGY

The most advanced nanotechnology projects related to energy are: Storage, conversion, manufacturing improvements by reducing materials and process rates, energy saving (by better thermal insulation for example), and enhanced renewable energy sources. A reduction of energy consumption can be reached by better insulation systems, by the use of more efficient lighting or combustion systems, and by use of lighter and stronger materials in the transportation sector. Currently used light bulbs only convert approximately 5% of the electrical energy into light. Nanotechnological approaches like light-emitting diodes (LEDs) or quantum caged atoms (QCAs) could lead to a strong reduction of energy consumption for illumination. Today's best solar cells have layers of several different semiconductors stacked together to absorb light at different energies but they still only manage to use 40 per cent of the Sun's energy. Commercially available solar cells have much lower efficiencies (15-20%).

Nanotechnology could help increase the efficiency of light conversion by using nanostructures with a continuum of bandgaps. The degree of efficiency of the internal combustion engine is about 30-40% at the moment. Nanotechnology could improve combustion by designing specific catalysts with maximized surface area. In 2005, scientists at the University of Toronto developed a spray-on nanoparticle substance that, when applied to a surface, instantly transforms it into a solar collector. An example for an environmentally friendly form of energy is the use of fuel cells powered by hydrogen, which is ideally produced by renewable energies. Probably the most prominent nanostructured material in fuel cells is the catalyst consisting of carbon supported noble metal particles with diameters of 1-5 nm.

Suitable materials for hydrogen storage contain a large number of small nanosized pores. Therefore many nanostructured materials like nanotubes, zeolites or alanates are under investigation. Nanotechnology can contribute to the further reduction of combustion engine pollutants by nanoporous filters, which can clean the exhaust mechanically, by catalytic converters based on nanoscale noble metal particles or by catalytic coatings on cylinder walls and catalytic nanoparticles as additive for fuels.

Because of the relatively low energy density of batteries the operating time is limited and a replacement or recharging is needed. The huge number of spent batteries and accumulators represent a disposal problem. The use of batteries with higher energy content or the use of rechargeable batteries or supercapacitors with higher rate of recharging using nanomaterials could be helpful for the battery disposal problem.

INFORMATION AND COMMUNICATION

Current high-technology production processes are based on traditional top down strategies, where nanotechnology has already been introduced silently. The critical length scale of integrated circuits is already at the nanoscale (50 nm and below) regarding the gate length of transistors in CPUs or DRAM devices. Electronic memory designs in the past have largely relied on the formation of transistors however research into crossbar switch based electronics have offered an alternative using reconfigurable interconnections between vertical and horizontal wiring arrays to create ultra high density memories.

Two leaders in this area are Nantero which has developed a carbon nanotube based crossbar memory called Nano-RAM and Hewlett-Packard which has proposed the use of memristor material as a future replacement of Flash memory. An example of such novel devices is based on spintronics.The dependence of the resistance of a material (due to the spin of the electrons) on an external field is called magnetoresistance. This effect can be significantly amplified (GMR-Giant Magneto-Resistance) for nanosized objects, for example when two ferromagnetic layers are separated by a nonmagnetic layer, which is several nanometers thick (*e.g.* Co-Cu-Co). The GMR effect has led to a strong increase in the data storage density of hard disks and made the gigabyte range possible. The so called tunneling magnetoresistance (TMR) is very similar to GMR and based on the spin dependent tunneling of electrons through adjacent ferromagnetic layers.

Both GMR and TMR effects can be used to create a non-volatile main memory for computers, such as the so called magnetic random access memory or MRAM. In 1999, the ultimate CMOS transistor developed at the Laboratory for Electronics and Information Technology in Grenoble, France, tested the limits of the principles of the MOSFET transistor with a diameter of 18 nm (approximately 70 atoms placed side by side). This was almost one tenth the size of the smallest industrial transistor in 2003. It enabled the theoretical integration of seven billion junctions on a €1 coin. However, the CMOS transistor, which was created in 1999, was not a simple research experiment to study how CMOS technology functions, but rather a demonstration of how this technology functions now that we ourselves are getting ever closer to working on a molecular scale.

Today it would be impossible to master the coordinated assembly of a large number of these transistors on a circuit and it would also be impossible to create this on an industrial level. In the modern communication technology traditional analog electrical devices are increasingly replaced by optical or optoelectronic devices due to their enormous bandwidth and capacity, respectively. Two promising examples are photonic crystals and quantum dots. Photonic crystals are materials with a periodic variation in the refractive index with a lattice constant that is half the wavelength of the light used. They offer a selectable band gap for the propagation of a certain wavelength, thus they resemble a semiconductor, but for light or photons instead of electrons. Quantum dots are nanoscaled objects, which can be used, among many other things, for the construction of lasers. The advantage of a quantum dot laser over the traditional semiconductor laser is that their emitted wavelength depends on the diameter of the dot. Quantum dot lasers are cheaper and offer a higher beam quality than conventional laser diodes.

The production of displays with low energy consumption could be accomplished using carbon nanotubes (CNT). Carbon nanotubes be electrically conductive and due to their small diameter of several nanometers, they can be used as field emitters with extremely high efficiency for field emission displays (FED). The principle of operation resembles that of the cathode ray tube, but on a much smaller length scale.

Entirely new approaches for computing exploit the laws of quantum mechanics for novel quantum computers, which enable the use of fast quantum algorithms. The Quantum computer will have quantum bit memory space termed qubit for several computations at the same time.

HEAVY INDUSTRY

Aerospace

Lighter and stronger materials will be of immense use to aircraft manufacturers, leading to increased performance. Spacecraft will also benefit, where weight is a major factor. Nanotechnology thereby helps to reduce the size of equipments used. Hang gliders halve their weight while increasing their strength and toughness through the use of nanotech materials. Nanotech is lowering the mass of supercapacitors that will increasingly be used to give power to assistive electrical motors for launching hang gliders off flatland to thermal-chasing altitudes.

Using nanotech applications, refineries producing materials such as steel and aluminium will be able to remove any impurities in the materials they create. Much like aerospace, lighter and stronger materials will be useful for creating vehicles that are both faster and safer. Combustion engines will also benefit from parts that are more hard-wearing and more heat-resistant.

GOODS

Nanotechnology is already impacting the field of consumer goods, providing products with novel functions ranging from easy-to-clean to scratch-resistant. Modern textiles are wrinkle-resistant and stain-repellent; in the mid-term clothes will become "smart", through embedded "wearable electronics". Already in use are different nanoparticle improved products. Especially in the field of cosmetics, such novel products have a promising potential. Nanotechnology can be applied in the production, processing, safety and packaging of food. A nanocomposite coating process could improve food packaging by placing anti-microbial agents directly on the surface of the coated film. Nanocomposites could increase or decrease gas permeability of different fillers as is needed for different products. They can also improve the mechanical and heat-resistance properties and lower the oxygen transmission rate. Research is being performed to apply nanotechnology to the detection of chemical and biological substances for sensing biochemical changes in foods.

The most prominent application of nanotechnology in the household is self-cleaning or "easy-to-clean" surfaces on ceramics or glasses. Nanoceramic particles have improved the smoothness and heat resistance of common household equipment such as the flat iron. The first sunglasses using protective and antireflective ultrathin polymer coatings are on the market. For optics, nanotechnology also offers scratch resistant surface coatings based on nanocomposites. Nano-optics could allow for an increase in precision of pupil repair and other types of laser eye surgery.

The use of engineered nanofibres already makes clothes water- and stain-repellent or wrinkle-free. Textiles with a nanotechnological finish can be washed less frequently and at lower temperatures. Nanotechnology has been used to integrate tiny carbon particles membrane and guarantee full-surface protection from electrostatic charges for the wearer. One field of application is in sunscreens. The traditional chemical UV protection approach suffers from its poor long-term stability. A sunscreen based on mineral nanoparticles such as titanium dioxide offer several advantages. Titanium oxide nanoparticles have a comparable UV protection property as the bulk material, but lose the cosmetically undesirable whitening as the particle size is decreased.

NANOTUBE

In fact, nanotubes long, narrow shape make them look like miniature needles, so it makes sense that they can function like a needle at the cellular level. Medical researchers are using this property by attaching molecules that are attracted to cancer cells to nanotubes to deliver drugs directly to the diseased cells. Another interesting property of nanotubes is that their electrical resistance changes significantly when other molecules attach themselves to the carbon atoms. Companies are using this property to develop sensors that can detect chemical vapours such as carbon monoxide or biological molecules. These are just a few of the potential uses of carbon nanotubes The following survey of carbon nanotube applications introduces these and many other uses.

Researchers and companies are working to use carbon nanotubes in various fields. The list below introduces many of these uses. Strong, lightweight composites of carbon nanotubes and other materials that can be used to build lightweight spacecraft. Cables made from carbon nanotubes strong enough to be used for the Space Elevator to drastically reduce the cost of lifting people and materials into orbit. Taking advantage of nanotubes ability to enter cancer cells by attaching targeting molecules which have an affinity to cancer cells as well as anti-cancer drugs to the nanotubes which safety transports an anti-cancer drug through the bloodstream to the tumor.

Stronger bicycle components made by adding carbon nanotubes to a matrix of carbon fibres. Improve the healing process for broken bones by providing a carbon nanotube scaffold for new bone material to grow on. Sensors using carbon nanotube detection elements capable of detecting a range of chemical vapours. These sensors depend upon the fact that the resistance of a carbon nanotube changes in the presence of a chemical vapour.

IMPLICATIONS

Due to the far-ranging claims that have been made about potential applications of nanotechnology, a number of serious concerns have been raised about what effects these will have on our society if realised, and what action if any is appropriate to mitigate these risks. There are possible dangers that arise with the development of nanotechnology. The Centre for Responsible Nanotechnology suggests that new developments could result, among other things, in untraceable weapons of mass destruction, networked cameras for use by the government, and weapons developments fast enough to destabilize arms races ("Nanotechnology Basics").

One area of concern is the effect that industrial-scale manufacturing and use of nanomaterials would have on human health and the environment, as suggested by nanotoxicology research. Groups such as the Centre for Responsible Nanotechnology have advocated that nanotechnology should be specially regulated by governments for these reasons. Others counter that overregulation would stifle scientific research and the development of innovations which could greatly benefit mankind.

Other experts, including director of the Woodrow Wilson Center's Project on Emerging Nanotechnologies David Rejeski, have testified that successful commercialization depends on adequate oversight, risk research strategy, and public engagement. More recently local municipalities have passed (Berkeley, CA) or are considering (Cambridge, MA)-ordinances requiring nanomaterial manufacturers to disclose the known risks of their products. The National Institute for Occupational Safety and Health is conducting research on how nanoparticles interact with the body's systems and how workers might be exposed to nano-sized particles in the manufacturing or industrial use of nanomaterials. NIOSH offers interim guidelines for working with nanomaterials consistent with the best scientific knowledge.

In "The Consumer Product Safety Commission and Nanotechnology," E. Marla Felcher suggests that the Consumer Product Safety Commission, which is charged with protecting the public against unreasonable risks of injury or death associated with consumer products, is ill-equipped to oversee the safety of complex, high-tech products made using nanotechnology. Longer-term concerns centre on the implications that new technologies will have for society at large, and whether these could possibly lead to either a post scarcity economy, or alternatively exacerbate the wealth gap between developed and developing nations. The effects of nanotechnology on the society as a whole, on human health and the environment, on trade, on security, on food systems and even on the definition of "human", have not been characterized or politicized.

Health and Environmental Concerns

Some of the recently developed *nanoparticle* products may have unintended consequences. Researchers have discovered that silver nanoparticles used in socks to reduce foot odour are being released in the wash with possible negative consequences. Silver nanoparticles, which are bacteriostatic, may then destroy beneficial bacteria which are important for breaking down organic matter in waste treatment plants or farms.

A study at the University of Rochester found that when rats breathed in nanoparticles, the particles settled in the brain and lungs, which led to significant increases in biomarkers for inflammation and stress response. A major study published more recently in Nature nanotechnology suggests some forms of carbon nanotubes – a poster child for the "nanotechnology revolution" – could be as harmful as asbestos if inhaled in sufficient quantities. Anthony Seaton of the Institute of Occupational Medicine in Edinburgh, Scotland, who contributed to the article on carbon nanotubes said "We know that some of them probably have the potential to cause mesothelioma. So those sorts of materials need to be handled very carefully." . In the absence of specific nano-regulation forthcoming from governments, Paull and Lyons (2008) have called for an exclusion of engineered nanoparticles from organic food.

Regulation

Calls for tighter regulation of nanotechnology have occurred alongside a growing debate related to the human health and safety risks associated with nanotechnology. Further, there is significant debate about who is responsible for the regulation of nanotechnology.

While some non-nanotechnology specific regulatory agencies currently cover some products and processes (to varying degrees) – by "bolting on" nanotechnology to existing regulations – there are clear gaps in these regimes. In "Nanotechnology Oversight: An Agenda for the Next Administration," former EPA deputy administrator J.

Clarence (Terry) Davies lays out a clear regulatory

roadmap for the next presidential administration and describes the immediate and longer term steps necessary to deal with the current shortcomings of nanotechnology oversight. Stakeholders concerned by the lack of a regulatory framework to assess and control risks associated with the release of nanoparticles and nanotubes have drawn parallels with bovine spongiform encephalopathy ('mad cow's disease), thalidomide, genetically modified food, nuclear energy, reproductive technologies, biotechnology, and asbestosis.

The Royal Society report identified a risk of nanoparticles or nanotubes being released during disposal, destruction and recycling, and recommended that "manufacturers of products that fall under extended producer responsibility regimes such as end-of-life regulations publish procedures outlining how these materials will be managed to minimize possible human and environmental exposure" (p.xiii). Reflecting the challenges for ensuring responsible life cycle regulation, the Institute for Food and Agricultural Standards has proposed standards for nanotechnology research and development should be integrated across consumer, worker and environmental standards. They also propose that NGOs and other citizen groups play a meaningful role in the development of these standards. The term "manufacturers" includes persons and businesses that produce nanotubes in California, or import carbon nanotubes into California for sale.

FABRICATION

Computer chips (and the silicon based transistors within them) are rapidly shrinking according to a predictable formula. According to the Semiconductor Industry Association's extrapolation of formulas such as this one (SIA roadmap), it is expected that the sizes of circuits within our chips will reach the size of only a few atoms in about 20 years.

Since almost all of our modern computers are made from silicon semiconductor transistors patterned and carved by light (photolithography), the shrinking of circuits predicted by the SIA may not be the most economical method for the future. An enormous amount of money has been invested in the semiconductor industry in order to consistently shrink and improve our semiconductor electronics.

Smaller circuits require less energy, operate more quickly and, of course, take up less space. Thus, Moore's law has been adhered to since computers first became commercially available. However, this simple shrinking of components can not continue for much longer. As transistors such as the Metal-Oxide Semiconductor Field Effect Transistor is made smaller, both its properties and manufacturing expense change with the scale. Currently, Ultraviolet light is used to create the silicon circuits with a lateral resolution around 200 nm (the wavelength of ultraviolet light). As the circuits shrink below 100 nm new fabrication methods must be created, resulting in increasing costs.

Furthermore, once the circuit size reaches only a few nanometers, quantum effects such as tunneling begin to become important, which drastically changes the ability for the computers to function normally. Thus, novel methods for computer chip fabrication have been and are being intensely sought by microchip manufactures.

MOLECULAR AND QUANTUM COMPUTING ARCITECTURES FOR NANOCOMPUTING

Quantum computing seeks to write, process and read information on the quantum level. It is at the nanoscale that quantum mechanical effects such as (the wave particle duality) begin to become apparent. Numerous scientists are seeking ways to store information within the quantum mechanical realm. This is not a simple task because of the delicate nature of quantum mechanical systems.

However, since the laws of quantum mechanics involves unintuitive principles such as superposition and entanglement, a quantum computer would be able to violate some rules that limit our classical computers. For instance, taking advantage of superposition would mean that a quantum bit of information, termed a qubit would be able to be used in several computations at the same time.

Taking advantage of entanglement would mean that the

information could be processed over long distances without the classical requirement of wires. Molecular computation is another method complimentary to quantum computing that seeks to write, process and read information within single molecules. One molecule that has proved most promising for molecular computation is Deoxyribonucleic acid (DNA).

DNA is a long polymer made of 4 different nucleotides that can be represented by the letters A, T, C and G. The order or sequence of these nucleotides within DNA provides the information for making protein, the main components of the molecular scale machinery used b4y living organisms to carry out life sustaining functions.

Mathematicians have figured out numerous ways to use DNA an the various proteins that come with it to carry out numerical computations that are notoriously difficult for silicon computers, namely NP-complete problems. The advantage that molecular computing using DNA has over conventional computing is that it is massively parallel. This means that each DNA molecule can function as a single processor, which greatly improves the speed of computation for complex problems.

MEDICAL

The other field in which molecular scale manipulation of matter is receiving abundant attention is medicine. Since all living organisms are composed of molecules, molecular biology has become the primary focus of biotechnology. Countless diseases have been cured by our ability to synthesize small molecules commonly know as 'drugs' that interact with the protein molecules that make up the molecular machinery that keeps us alive.

Our understanding of how proteins interact with DNA, phospholipids and other biological molecules is what allows such progress. Living systems are able to live because of the vast amount of highly ordered molecular machinery from which they are built. The central dogma of molecular biology states that the information required to build a living cell or organism is stored in the DNA (which was described above for its use in molecular computation). This information is transferred from the DNA to the proteins by the processes called transcription and translation. These processes are all executed by various biomolecular components, mostly protein and nucleic acids. Molecular biology is a field in which the study of these interactions has led to the discovery of numerous pharmaceuticals that have been enormously effective in curing disease.

Understanding of molecular mechanisms including substrate recognition, energy expenditure, electron transport, membrane activity and much more have greatly improved our medical technology. So, what does this have to do with nanotechnology? First of all it shows the abilities of molecular scale machinery. Since the goal of nanotechnology is molecular and atomic precision, nanotechnology has much (if not everything) to learn from nature. Copying, borrowing and learning tricks from nature is one of the primary techniques used by nanotechnology and has been termed biomimetics.

2

Universal Computing

The concept of a "Universal Constructor" is perhaps more recent. The concept was well understood by von Neumann in the 1940's, who defined a "Universal Constructor" in a two-dimensional cellular automata world. Such a model is a mathematical abstraction something like an infinite checkerboard, with several different types of "checkers" that might be on each square. The different pieces spontanesouly change and move about depending on what pieces occupy neighbouring squares, in accordance with a pre-defined set of rules.

Von Neumann used the concept of a Universal Constructor in conjunction with a Universal Computer as the core components in a self-replicating system. The possibility of fabricating structures by putting "...the atoms down where the chemist says..." was recognized by Feynman in 1959. Drexler recognized the value of the "assembler" in 1977.

The assembler is analogous to von Neumann's Universal Constructor, but operates in the normal three dimensional world and can build a large, atomically precise structures by manipulating atoms and small clusters of atoms. He published the concept in 1981 and in subsequent work has proposed increasingly detailed designs for such devices.

The basic design of Drexler's assembler consists of:

- A molecular computer,
- One or more molecular positioning devices (which might resemble very small robotic arms), and
- A well defined set of chemical reactions (perhaps one or two dozen) that take place at the tip of the arm

and are able to fabricate a wide range of structures using site-specific chemical reactions.

It is now common for forecasts of future technical abilities to include the ability to fabricate molecular devices and molecular machines with atomic precision. While there continues to be debate about the exact time frame, it is becoming increasingly accepted that we will, eventually, develop the ability to economically fabricate a truly wide range of structures with atomic precision.

This will be of major economic value. Most obviously a molecular manufacturing capability will be a prerequisite to the construction of molecular logic devices. The continuation of present trends in computer hardware depends on the ability to fabricate ever smaller and ever more precise logic devices at ever decreasing costs.

The limit of this trend is the ability to fabricate molecular logic devices and to connect them in complex patterns at the molecular level. The manufacturing technology needed will, almost of necessity, be able to economically manufacture large structures (computers) with atomic precision (molecular logic elements). This capability will also permit the economical manufacture of materials with properties that border on the limits imposed by natural law.

The strength of materials, in particular, will approach or even exceed that of diamond. Given the broad range of manufactured products that devote substantial mass to load-bearing members, such a development by itself will have a significant impact. A broad range of other manufactured products will also benefit from a manufacturing process that offers atomic precision at low cost.

Given the promise of such remarkably high payoffs it is natural to ask exactly what such systems will look like, exactly how they will work, and exactly how we will go about building them. One might also enquire as to the reasons for confidence that such an enterprise is feasible, and why one should further expect that our current understanding of chemistry and physics (embodied in a number of computational chemistry packages) should be sufficient to explain the operating principles of such systems. It is here that the value of computational nanotechnology can be most clearly seen. Molecular machine proposals, provided that they are specified in atomic detail (and are of a size that can be dealt with by current software and hardware), can be modeled using the tools of computational chemistry.

There are two modeling techniques of particular utility. The first is molecular mechanics, which utilizes empirical force fields to model the forces acting between nuclei. The second is higher order ab initio calculations, which will be discussed in a few paragraphs.

MOLECULAR MECHANICS

More complex analyses, particularly analyses that involve searching through large configuration spaces, can limit the size of system that can be effectively handled. As will be discussed later the need to search through large configuration spaces (as when determining the native folded structure of an arbitrary protein) can be avoided by the use of relatively rigid structures (which differ from relatively floppy proteins and have few possible configurations).

The modeling of machine components in vacuum reduces the need to model solvation effects, which can also involve significant computational effort. In molecular mechanics the individual nuclei are usually treated as point masses. While quantum mechanics dictates that there must be a certain degree of positional uncertainty associated with each nucleus, this positional uncertainty is normally significantly smaller than the typical internuclear distance. Bowen and Allinger provide a good recent overview of the subject.

While the nuclei can reasonably be approximated as point masses, the electron cloud must be dealt with in quantum mechanical terms. However, if we are content to know only the positions of the nuclei and are willing to forego a detailed understanding of the electronic structure, then we can effectively eliminate the quantum mechanics.

For example, the H_2 molecule involves two nuclei. While it would be possible to solve Schrodinger's equation to determine the wave function for the electrons, if we are content simply to know the potential energy contributed by the electrons (and do not enquire about the electron distribution) then we need only know the electronic energy as a function of the distance between the nuclei.

That is, in many systems the only significant impact that the electrons have on nuclear position is to make a contribution to the potential energy E of the system. In the case of H_2, E is a simple function of the internuclear distance r. The function E(r) summarizes and replaces the more complex and more difficult to determine wave function for the electrons, as well as taking into account the inter-nuclear repulsion and the interactions between the electrons and the nuclei. The two hydrogen nuclei will adopt a position that minimizes E(r). As r becomes larger, the potential energy of the system increases and the nuclei experience a restoring force that returns them to their original distance. Similarly, as r becomes smaller and the two nuclei are pushed closer together, we also find that a restoring force pushes them farther apart, again restoring them to an equilibrium distance.

More generally, if we know the positions r_1, r_2.... r_N of N nuclei, then E(r_1, r_2.... r_N) gives the potential energy of the system. Knowing the potential energy as a function of the nuclear positions, we can readily determine the forces acting on the individual nuclei and therefore can compute the evolution of their position over time. The function E is a newtonian potential energy function (not quantum mechanical), despite the fact that the particular value of E at a particular point could be computed from Schrodinger's equation. That is, the potential energy E is a newtonian concept, but the particular values of E at particular points are determined by Schrodinger's equation.

While it would in principle be possible to determine E by solving Schrodinger's equation, in practice it is usual to use available experimental data and to infer the nature and shape of E by interpolation. This approach, in which empirically derived potential energy functions are created by interpolation from experimental data, has spawned a wide range of potential energy functions, many of which are sold commercially. Because the gradient of the potential energy function E defines a conservative force field F, molecular mechanics methods are also called "force field" methods.

While it is common to refer to "empirical force field" methods, the more recent use of ab initio methods to provide data points to aid in the design of the force fields makes this term somewhat inaccurate, though still widely used. The utility of molecular mechanics depends crucially on the development of accurate force fields. Good quality force fields have been developed for a fairly broad range of compounds including many compounds of interest in biochemistry.

While we will not attempt to survey the wide range of force fields that are available, one particular subset of compounds for which good quality force fields are available involve H, C, N, O, F, Si, P, S, Cl when they are restricted to form chemically uncomplicated structures (*e.g.*, bond strain is not too great, dangling bonds are few or absent, etc). Many atomically precise structures which should be useful in nanotechnology fall in this class and can be modeled with an accuracy adequate to determine the behaviour of molecular machines. The, given essentially arbitrary coordinates of the carbon and hydrogen atoms in a system, Brenner's potential will return the energy of the system. This permits molecular dynamical modeling of arbitrary hydrocarbon systems, including systems which use synthetic reactions involved in the synthesis of diamond.

A particular reaction of interest is the selective abstraction of a chosen hydrogen atom from a diamond surface. See *Surface patterning by atomically-controlled chemical forces: molecular dynamics simulations* by Sinnott *et al.*, Surface Science 316 (1994), L1055-L1060; see also the work of Robertson *et al.* for an illustration of the use of Brenner's potential in modeling the behaviour of graphitic gears, including failure modes).

Brenner has commented on the utility of this potential for modeling proposed molecular machine systems. Of course, the "accuracy" of the force fields depends on the application. A force field which was accurate to (say) 10 kcal/mole would be unable to correctly predict many properties of interest in biochemistry.

For example, such a force field would lead to serious errors in predicting the correct three-dimensional structure of a protein. Given the linear sequence of amino acids in a protein, the protein folding problem is to determine how it will fold in three dimensions when put in solution. Often, the correct configuration will have an energy that differs from other (incorrect) configurations by a relatively modest amount, and so a force field of high accuracy is required. Unlike the protein folding problem, where an astronomical range of configurations of similar energy are feasible, the bearing basically has only one configuration: A bearing.

While the protein has many unconstrained torsions, there are no unconstrained torsions in the bearing. To significantly change any torsion angle in the bearing would involve ripping apart bonds. Thus, the same force field which is of marginal utility in dealing with strands of floppy protein is quite adequate for solid blocks of stiff diamondoid material. Not only will small errors in the force field still result in an accurate prediction of the global minima, (which in this case will be a single large basin in the potential energy surface) but also the range of possible structures is so sharply limited that little or no computational effort need be spent comparing the energies of different configurations. Long computational runs to evaluate the statistical properties of ensembles of configurations are thus eliminated.

This style of design has been called, only half in jest, "molecular bridge building" because bridges are also designed with large safety margins. This observation, that the same force field that is inadequate for one class of structures is quite adequate for the design and modeling of a different class of structures, leads to a more general principle.

Computational experiments generally provide an answer with some error distribution. If the errors produced by the model are of a similar size to the errors that would result in incorrect device function, then the model is unreliable. On the other hand, if the errors in the model are small compared with the errors that will produce incorrect device function, then the results of the model are likely to be reliable. If a device design falls in the former category, *i.e.,* small errors in the model will produce conflicting forecasts about device function, then the conservative course of action is to reject the proposed design and keep looking. An even stronger (although somewhat more subtle) statement is possible. The bearing illustrated in figures is simply a single bearing from a very large class of bearings. The strain in the axle and the sleeve is proportional to the diameter of the bearing. By increasing the diameter, we can reduce the strain. Thus, we can design bearings in this broad class in which the strain can be reduced to whatever level we desire. Because we are dealing with what amounts to a strained block of diamond, the fact that we can reduce strain arbitrarily means that we can design a bearing whose local structure bears as close a resemblance to an unstrained block of diamond as might be wished.

We can therefore be very confident indeed that some member of this class will perform the desired function (that of a bearing) and will work in accordance with our expectations. Given that we have developed software tools that are capable of creating and modeling most of the members of a broad class of devices, then we can investigate many individual class members. This investigation then lets us make rather confident statements about the functionality that members of this class can provide, even if there might be residual doubts about individual class members.

A moments reflection will show that the class of objects which are chemically reasonably inert, relatively stiff (no free torsions), and which interact via simple repulsive forces that occur on contact; can describe a truly vast class of machines. Indeed, it is possible to design computers, robotic arms and a wide range of other devices using molecular parts drawn from this class.

FORCE FIELD

While empirical force fields are sufficiently accurate to model the behaviour of chemically stable stiff structures interacting with other chemically stable stiff structures, they do not (at present) provide sufficient accuracy to deal with chemical transitions. Thus, if we wish to model the manufacture of a molecular part then we must use higher order ab initio techniques.

These techniques impose severe constraints on the number of atoms that can be modeled (perhaps one or two dozen heavy atoms, depending on the hardware, software, and specific type of modeling being attempted), but can provide an accuracy sufficient to analyse the chemical reactions that must necessarily take place during the synthesis of large, atomically precise structures.

An analysis of the abstraction of a hydrogen atom from various structures, including isobutane (which serves as a model of the diamond (111) surface), has been carried out to illustrate the kind of reactions that are of interest. More generally, higher order ab initio techniques are sufficient to analyse the addition or removal of a small number of atoms from a specific site on a work piece.

Synthesis of a large object would then consist of repeated site specific applications of a small number of basic operations, where each basic operation changed the chemical structure of only a small number of atoms at a time. Provided that we reject reaction mechanisms where the result predicted depends on errors that are smaller than can reasonably be modeled, the analysis of these basic operations can be satisfactorily carried out with current methods and hardware.

CURRENT MODELING METHODS

Clearly, not all molecular machines satisfy these constraints. Is the range of molecular machines which do satisfy these constraints sufficiently large to justify the effort of designing and modeling them? And in particular, can we satisfactorily model The fundamental purpose of an assembler is to position atoms. To this end, it is imperative that we have models that let us determine atomic positions, and this is precisely what molecular mechanics provides. Robotic arms or other positioning devices are basically mechanical in nature, and will allow us to position molecular parts during the assembly process.

Molecular mechanics provides us with an excellent tool for modeling the behaviour of such devices. The second fundamental requirement is the ability to make and break bonds at specific sites. While molecular mechanics provides an excellent tool for telling us where the tip of the assembler arm is located, current force fields are not adequate to model the specific chemical reactions that must then take place at the tip/work-piece interface involved in building an atomically precise part (though this statement must be modified in light of the work done with Brenner's potential, discussed above). For this, higher order ab initio calculations are sufficient.

The glaring omission in this discussion is the modeling of the kind of electronic behaviour that occurs in switching devices. Clearly, it is possible to model electronic behaviour with some degree of accuracy, and equally clearly molecular machines that are basically electronic in nature will be extremely useful. It will therefore be desirable to extend the range of computational models discussed here to include such devices. For the moment, however, the relatively modest inclusion of electrostatic motors as a power source is probably sufficient to provide us with adequate "design space" to design and model an assembler.

While it might at first glance appear that electronics will be required for the computational element in the assembler, in fact molecular mechanical logic elements will be sufficient. Babbage's proposal from the 1800's clearly demonstrates the feasibility of mechanical computation (it is interesting to note that a working model of Babbage's difference engine has been built and is on display at the British Museum of Science.

They were careful to use parts machined no more accurately than the parts available to Babbage, in order to demonstrate that his ideas could have been implemented in the 1800's). Drexler's analysis of a specific molecular mechanical logic element and further analysis of system design issues also make it clear that molecular mechanical computation is sufficient for the molecular computer required in an assembler. Molecular electronic proposals for computation have not been worked out as clearly at the system level as the molecular mechanical concepts. Thus, at the moment, molecular mechanical proposals are better understood, at least in this particular context. This situation is likely to change, and when it does the electronic designs could be incorporated (where appropriate) into the design and modeling of an assembler. However, it is not entirely obvious that electronic designs will prove superior to molecular mechanical designs, particularly when device parameters such as size and energy dissipation are considered.

While it seems virtually certain that electronic devices will prove faster than molecular mechanical devices, it is less obvious that electronic devices must necessarily be either smaller or dissipate less energy (though confer more recent work on reversible logic). Indeed, given the difficulty of localizing individual electrons, it seems possible that electronic devices will prove to be inherently larger than molecular mechanical devices. This might provide a long term role for molecular mechanical devices as high density memory elements, or as logic elements when space (or atom count) is particularly constrained.

Whether the assembler is designed and built with an electronic or mechanical computer is less significant than designing and building it. By way of example, a stored programme computer can be built in many different ways. Vacuum tubes, transistors, moving mechanical parts, fluidics, and other methods are all entirely feasible.

Delaying the development of the Eniac because transistors are better than vacuum tubes would have been a most unwise course of action. Similarly, as we consider the design, modeling, and construction of an assembler, we should not hesitate to use simpler methods that we can understand in favour of better methods that are not yet fully in hand.

The methods of computational chemistry available today allow us to model a wide range of molecular machines with an accuracy sufficient in many cases to determine how well they will work. We can deliberately confine ourselves to the subset of devices where our modeling tools are good enough to give us confidence that the proposals should work. This range includes bearings, computers, robotic arms, site-specific chemical reactions utilized in the synthesis of complex structures, and more complex systems built up from these and related components. Drexler's assembler and related devices can be modeled using our current approaches and methods.

MOLECULAR COMPILERS

Computational nanotechnology includes not only the tools and techniques required to model proposed molecular machines, it must also include the tools required to specify such machines. Molecular machine proposals that would require millions or even billions of atoms have been made.

The total atom count of an assember might be roughly a billion atoms. While commerically available molecular modeling packages provide facilities to specify arbitrary structures, it is usually necessary to "point and click" for each atom involved. This is obviously unattractive for a device as complex as an assembler with its roughly one billion atoms. It should be clear that molecular CAD tools will be required if we are to specify such complex structures in atomic detail with a reasonable amount of effort.

An essential development is that of "molecular compilers" which accept, as input, a high level description of an object and produce, as output, the atomic coordinates, atom types, and bonding structure of the object.

Using this approach, it will be possible to substantially reduce the development time for complex molecular machines, including Drexler's assembler. This approach is similar in spirit to the computer aided design and modeling used to speed the development of many products today. The author was part of a general purpose computer start up which successfully designed and built a new computer from scratch. This included the hardware, software, compilers, operating system, etc.

During this process, extensive use was made of computational models to verify each level of the design. The operating system was written in Pascal, and checked out on another computer. The compilers were written in Pascal, and also checked out on another computer. The code produced by the compilers was checked out on an instruction set simulator. The microcode was checked out on a micro-instruction simulator. The logic design was checked out with logic level simulation tools, and circuit simulation packages were used to verify the detailed electronic design. All this work was done at the same time.

The software was written and debugged even though the machine on which it would eventually be executed didn't exist. The microcode was written and debugged before the hardware was available. When the hardware was finally made available, system integration went relatively rapidly. Imagine, for a moment, how long it would have taken to develop this system had we carried out the development in the obvious sequential fashion. First, we would have implemented the hardware and only then begun work on the microengine.

Later, with the hardware and microengine fully checked out and working, we could have developed and debugged the microcode. With this firmly in hand, we could then have written the compilers and verified the code they produced. Finally, we could have implemented and checked out the operating system. Needless to say, such a strategy would have been very slow and tedious.

Doing things in the simple and most obvious way often takes a lot longer than is needed. If we were to approach the design and construction of an assembler using the simple serial method, it would take a great deal longer than if we systematically attacked and simultanesouly solved the problems that arise at all levels of the design at one and the same time. That is, by using methods similar to those used to design a modern computer, including intensive computational modeling of individual components and sub- systems, we can greatly shorten the time required to design and build complex molecular machines.

This can be further illustrated by considering the traditional manner of growth of our synthetic capability over time. Today, we find we are able to synthesize a certain range of compounds and structures. As time goes by, we will be able to synthesize an ever larger range of structures. This growth in our ability will proceed on a broad front, and reflects the efforts of a broad range of researchers who are each pursuing individual goals without concern about the larger picture into which they might fit. As illustrated in figure, given sufficient time we will eventually be able to synthesize complex molecular machines simply because we will eventually develop the ability to synthesize just about anything.

However, if we wish to develop a particular kind of device, *e.g.*, an assembler, then we can speed the process up by conducting computational experiments designed to clarify the objective and to specify more precisely the path from our current range of synthetic capabilities to the objective. Such computational experiments are inexpensive, can provide very detailed information, are possible for any structure (whether we can or cannot synthesize it) and will, in general, reduce the "time to market" for the selected product.

Computational experiments let us examine structures quickly and easily, rejecting those which have obvious defects (a precursor to the bearing shown in figure, for example, was too strained. By modifying the design and again minimizing the structure, we found a design with an acceptable strain). This kind of examination of the "design space" is impossible with physical experiments today, but is easily done with computational experiments.

Computational experiments also provide more information. For example, molecular dynamics can literally provide information about the position of each individual atom over time, information which would usually be inaccessible in a physical experiment. Of course, the major advantage of computational experiments over physical experiments in the current context is the simple fact that physical experiments aren't possible for molecular machines that we can't make with today's technology.

By using computational models derived from the wealth of experimental data that is available today, we can (within certain accuracy bounds) describe the behaviour of proposed systems that we plan to build in the future. If we deliberately design systems that are sufficiently robust that we are confident they will work regardless of the small errors that must be incurred in the modeling process, we can design systems today that we will not be able to build for some years, and yet still have reasonable confidence that they will work.

By fully utilizing the experience that has been developed in the rapid design and development of complex systems we can dramatically reduce the development time for molecular manufacturing systems. It is possible to debate how long it will be before we achieve a robust molecular manufacturing capability. However, it is very clear that we'll get there sooner if we develop and make intelligent use of molecular design tools and computational models. These will let us design and check the blueprints for the new molecular manufacturing technologies that we now see on the horizon, and will let us chart a more rapid and more certain path to their development.

NANOSENSOR

Nanosensors are any biological, chemical, or physical sensory points used to convey information about nanoparticles to the macroscopic world. Though humans have not yet been able to synthesize nanosensors, predictions for their use mainly include various medicinal purposes and as gateways to building other nanoproducts, such as computer chips that work at the nanoscale and nanorobots.

Presently, there are several ways proposed to make nanosensors, including top-down lithography, bottom-up assembly, and molecular self-assembly.

Medicinal uses of nanosensors mainly revolve around the potential of nanosensors to accurately identify particular cells or places in the body in need. By measuring changes in volume, concentration, displacement and velocity, gravitational, electrical, and magnetic forces, pressure, or temperature of cells in a body, nanosensors may be able to distinguish between and recognize certain cells, most notably those of cancer, at the molecular level in order to deliver medicine or monitor development to specific places in the body. In addition, they may be able to detect macroscopic variations from outside the body and communicate these changes to other nanoproducts working within the body. One example of nanosensors involves using the fluorescence properties of cadmium selenide quantum dots as sensors to uncover tumors within the body. By injecting a body with these quantum dots, a doctor could see where a tumor or cancer cell was by finding the injected quantum dots, an easy process because of their fluorescence.

Developed nanosensor quantum dots would be specifically constructed to find only the particular cell for which the body was at risk. A downside to the cadmium selenide dots, however, is that they are highly toxic to the body. As a result, researchers are working on developing alternate dots made out of a different, less toxic material while still retaining some of the fluorescence properties.

In particular, they have been investigating the particular benefits of zinc sulfide quantum dots which, though they are not quite as fluorescent as cadmium selenide, can be augmented with other metals including manganese and various lanthanide elements. In addition, these newer quantum dots become more fluorescent when they bond to their target cells. Potential predicted functions may also include sensors used to detect specific DNA in order to recognize explicit genetic defects, especially for individuals at high-risk and implanted sensors that can automatically detect glucose levels for diabetic subjects more simply than current detectors.

DNA can also serve as sacrificial layer for manufacturing CMOS IC, integrating a nanodevice with sensing capabilities. Therefore, using proteomic patterns and new hybrid materials, nanobiosensors can also be used to enable components configured into a hybrid semiconductor substrate as part of the circuit assembly. The development and miniaturization of nanobiosensors should provide interesting new opportunities. Other projected products most commonly involve using nanosensors to build smaller integrated circuits, as well as incorporating them into various other commodities made using other forms of nanotechnology for use in a variety of situations including transportation, communication, improvements in structural integrity, and robotics. Nanosensors may also eventually be valuable as more accurate monitors of material states for use in systems where size and weight are constrained, such as in satellites and other aeronautic machines.

Existing Nanosensors

Currently, the most common mass-produced functioning nanosensors exist in the biological world as natural receptors of outside stimulation. For instance, sense of smell, especially in animals in which it is particularly strong, such as dogs, functions using receptors that sense nanosized molecules.

Certain plants, too, use nanosensors to detect sunlight; various fish use nanosensors to detect minuscule vibrations in the surrounding water; and many insects detect sex pheromones using nanosensors. Certain electromagnetic sensors have also been in use in photoelectric systems. These work because the specific sensors called, aptly, photosensors are easily influenced by light of various wavelengths.

The electromagnetic source transfers energy to the photosensors and energizes them into an excited state which causes them to release an electron into a semiconductor. At that point, it is relatively easy to detect the electricity coming from the sensors, and thus easy to know if the sensors are receiving light. Though more advanced uses of photosensors incorporating other forms of nanotechnology have yet to be implemented into consumer society, most film cameras have used photosensors at the nano size for years. Traditional film uses a layer of silver ions that become excited by solar energy and clump into groups, as small as four atoms apiece in some cases, that scatter light and appear dark on the frame.

Various other types of film can be made using a similar process to detect other specific wavelengths of light, including x-rays, infrared, and ultraviolet. One of the first working examples of a synthetic nanosensor was built by researchers at the Georgia Institute of Technology in 1999. It involved attaching a single particle onto the end of a carbon nanotube and measuring the vibrational frequency of the nanotube both with and without the particle. The discrepancy between the two frequencies allowed the researchers to measure the mass of the attached particle.

Chemical sensors, too, have been built using nanotubes to detect various properties of gaseous molecules. Carbon nanotubes have been used to sense ionization of gaseous molecules while nanotubes made out of titanium have been employed to detect atmospheric concentrations of hydrogen at the molecular level. Many of these involve a system by which nanosensors are built to have a specific pocket for another molecule. When that particular molecule, and only that specific molecule, fits into the nanosensor, and light is shone upon the nanosensor, it will reflect different wavelengths of light and, thus, be a different colour.

Production Methods

There are currently several hypothesized ways to produce nanosensors. Top-down lithography is the manner in which most integrated circuits are now made. It involves starting out with a larger block of some material and carving out the desired form. These carved out devices, notably put to use in specific microelectromechanical systems used as microsensors, generally only reach the micro size, but the most recent of these have begun to incorporate nanosized components.

Another way to produce nanosensors is through the bottom-up method, which involves assembling the sensors out of even more minuscule components, most likely individual atoms or molecules. This would involve moving atoms of a particular substance one by one into particular positions which, though it has been achieved in laboratory tests using tools such as atomic force microscopes, is still a significant difficulty, especially to do en masse, both for logistic reasons as well as economic ones.

Most likely, this process would be used mainly for building starter molecules for self-assembling sensors. he third way, which promises far faster results, involves self-assembly, or "growing" particular nanostructures to be used as sensors. This most often entails one of two types of assembly. The first involves using a piece of some previously created or naturally formed nanostructure and immersing it in free atoms of its own kind. After a given period, the structure, having an irregular surface that would make it prone to attracting more molecules as a continuation of its current pattern, would capture some of the free atoms and continue to form more of itself to make larger components of nanosensors.

The second type of self-assembly starts with an already complete set of components that would automatically assemble themselves into a finished product. Though this has been so far successful only in assembling computer chips at the micro size, researchers hope to eventually be able to do it at the nanometer size for multiple products, including nanosensors. Accurately being able to reproduce this effect for a desired sensor in a laboratory would imply that scientists could manufacture nanosensors much more quickly and potentially far more cheaply by letting numerous molecules assemble themselves with little or no outside influence, rather than having to manually assemble each sensor.

NANOROBOTS

Nanorobotics is emerging as a demanding field dealing with miniscule things at molecular level. Nanorobots are quintessential nanoelectromechanical systems designed to perform a specific task with precision at nanoscale dimensions. Its advantage over conventional medicine lies on its size. Particle size has effect on serum lifetime and pattern of deposition. This allows drugs of nanosize to be used in lower concentration and has an earlier onset of therapeutic action.

It also provides materials for controlled drug delivery by directing carriers to a specific location. The typical medical nanodevice will probably be a micron-scale robot assembled from nanoscale parts. These nanorobots can work together in response to environment stimuli and programmed principles to produce macro scale results.

ELEMENTS OF NANOROBOTS

Carbon will likely be the principal element comprising the bulk of a medical nanorobot, probably in the form of diamond or diamondoid/fullerene nanocomposites. Many other light elements such as hydrogen, sulfur, oxygen, nitrogen, fluorine, silicon, etc. will be used for special purposes in nanoscale gears and other components. The chemical inertness of diamond is proved by several experimental studies. One such experiment conducted on mouse peritoneal macrophages cultured on DLC showed no significant excess release of lactate dehydrogenase or of the lysosomal enzyme beta N-acetyl-D-glucosaminidase an enzyme known to be released from macrophages during inflammation.

Morphological examination revealed no physical damage to either fibroblasts or macrophages, and human osteoblast like cells confirming the biochemical indication that there was no toxicity and that no inflammatory reaction was elicited in vitro. The smoother and more flawless the diamond surface, the lesser is the leukocyte activity and fibrinogen adsorption. Interestingly, on the rougher "polished" surface, a small number of spread and fused macrophages were present, indicating that some activation had occurred.

The exterior surface with near-nanometer smoothness results in very low bioactivity. Due to the extremely high surface energy of the passivated diamond surface and the strong hydrophobicity of the diamond surface, the diamond exterior is almost completely chemically inert.

DESIGN OF NANOROBOTS

Nanorobots will possess full panoply of autonomous subsystems whose design is derived from biological models. Drexler evidently was the first to point out, in 1981, that complex devices resemble biological models in their structural components. The various components in the nanorobot design may include onboard sensors, motors, manipulators, power supplies, and molecular computers.

Perhaps the best-known biological example of such molecular machinery is the ribosome the only freely programmable nanoscale assembler already in existence. The mechanism by which protein binds to the specific receptor site might be copied to construct the molecular robotic arm.

The manipulator arm can also be driven by a detailed sequence of control signals, just as the ribosome needs mRNA to guide its actions. These control signals are provided by external acoustic, electrical, or chemical signals that are received by the robot arm via an onboard sensor using a simple "broadcast architecture a technique which can also be used to import power. the biological cell may be regarded as an example of a broadcast architecture in which the nucleus of the cell send signals in the form of mRNA to ribosomes in order to manufacture cellular proteins.

Assemblers are molecular machine systems that could be described as systems capable of performing molecular manufacturing at the atomic scale which require control signals provided by an onboard nanocomputer This programmable nanocomputer must be able to accept stored instructions which are sequentially executed to direct the manipulator arm to place the correct moiety or nanopart in the desired position and orientation, thus giving precise control over the timing and locations of chemical reactions or assembly operations.

CONSTRUCTION

There are two main approaches to building at the nanometer scale: positional assembly and self-assembly. In positional assembly, investigators employ some devices such as the arm of a miniature robot or a microscopic set to pick up molecules one by one and assemble them manually. In contrast, self-assembly is much less painstaking, because it takes advantage of the natural tendency of certain molecules to seek one another out. With self-assembling components, all that investigators have to do is put billions of them into a beaker and let their natural affinities join them automatically into the desired configurations. Making complex nanorobotic systems requires manufacturing techniques that can build a molecular structure via computational models of diamond mechanosynthesis.

DMS is the controlled addition of carbon atoms to the growth surface of a diamond crystal lattice in a vacuum-manufacturing environment. Covalent chemical bonds are formed one by one as the result of positionally constrained mechanical forces applied at the tip of a scanning probe microscope apparatus, following a programmed sequence.

Recognition of Target Site by Nanorobots

Different molecule types are distinguished by a series of chemotactic sensors whose binding sites have a different affinity for each kind of molecule. The control system must ensure a suitable performance. It can be demonstrated with a determined number of nanorobots responding as fast as possible for a specific task based scenario. In the 3D workspace the target has surface chemicals allowing the nanorobots to detect and recognize it. Manufacturing better sensors and actuators with nanoscale sizes makes them find the source of release of the chemical. Nanorobot Control Design (NCD) simulator was developed, which is software for nanorobots in environments with fluids dominated by Brownian motion and viscous rather than inertial forces.

First, as a point of comparison, the scientists used the nanorobots' small Brownian motions to find the target by random search. In a second method, the nanorobots monitor for chemical concentration significantly above the background level. After detecting the signal, a nanorobot estimates the concentration gradient and moves towards higher concentrations until it reaches the target.

In the third approach, nanorobots at the target release another chemical, which others use as an additional guiding signal to the target. With these signal concentrations, only nanorobots passing within a few microns of the target are likely to detect the signal. Thus, we can improve the response by having the nanorobots maintain positions near the vessel wall instead of floating throughout the volume flow in the vessel from monitoring the concentration of a signal from others; a nanorobot can estimate the number of nanorobots at the target. So, the nanorobot uses this information to determine when enough nanorobots are at the target, thereby terminating any additional "attractant" signal a nanorobot may be releasing. It is found that the nanorobots stop attracting others once enough nanorobots have responded. The amount is considered enough when the target region is densely covered by nanorobots. Thus these tiny machines work at the target site accurately and precisely to that extent only to which it is designed to do.

Nanorobots Evading the Immune System

Every medical nanorobot placed inside the human body will encounter phagocytic cells many times during its mission. Thus all Nanorobots, which are of a size capable of ingestion by phagocytic cells, must incorporate physical mechanisms and operational protocols for avoiding and escaping from phagocytes. The initial strategy for medical nanorobots is first to avoid phagocytic contact or recognition.

To avoid being attacked by the host's immune system, the best choice is to have an exterior coating of passive diamond. The smoother and flawless the coating, the lesser is the reaction from the body's immune system. And if this fails then to avoid it's binding to the phagocyte surface that leads to phagocytic activation. If trapped, the medical nanorobot can induce exocytosis of the phagosomal vacuole in which it is lodged or inhibit both phagolysosomal fusion and phagosome metabolism. In rare circumstances, it may be necessary to kill the phagocyte or to blockade the entire phagocytic system.

The most direct approach for a fully functional medical nanorobot is to employ its motility mechanisms to locomote out of, or away from, the phagocytic cell that is attempting to engulf it. This may involve reverse cytopenetration, which must be done cautiously (*e.g.*, the rapid exit of nonenveloped viruses from cells can be cytotoxic). It is possible that frustrated phagocytosis may induce a localized compensatory granulomatous reaction.

Medical nanorobots therefore may also need to employ simple but active defensive strategies to forestall granuloma formation. Metabolizing local glucose and oxygen for energy can do the powering of the nanorobots. In a clinical environment, another option would be externally supplied acoustic energy. When the task of the nanorobots is completed, they can be retrieved by allowing them to exfuse themselves via the usual human excretory channels or can also be removed by active scavenger systems.

Nanorobots in Cancer Detection and Treatment

The development of nanorobots may provide remarkable advances for diagnosis and treatment of cancer. Nanorobots could be a very helpful and hopeful for the therapy of patients, since current treatments like radiation therapy and chemotherapy often end up destroying more healthy cells than cancerous ones. From this point of view, it provides a non-depressed therapy for cancer patients. The Nanorobots will be able to distinguish between different cell types that is the malignant and the normal cells by checking their surface antigens (they are different for each type of cell).

This is accomplished by the use of chemotactic sensors keyed to the specific antigens on the target cells. Another approach uses the innovative methodology to achieve decentralized control for a distributed collective action in the combat of cancer. Using chemical sensors they can be programmed to detect different levels of E-cadherin and beta-catenin in primary and metastatic phases. Medical nanorobots will then destroy these cells, and only these cells.

The following control methods were considered:

- *Random*: Nanorobots moving passively with the fluid reaching the target only if they bump into it due to Brownian motion.
- *Follow gradient*: Nanorobots monitor concentration intensity for E-cadherin signals, when detected, measure and follow the gradient until reaching the target. If the gradient estimate subsequent to signal detection finds no additional signal in50ms, the nanorobot considers the signal to be a false positive and continues flowing with the fluid.
- *Follow gradient with attractant*: As above, but

nanorobots arriving at the target, they release in addition a different chemical signal used by others to improve their ability to find the target. Thus, a higher gradient of signal intensity of E-cadherin is used as chemical parameter identification in guiding nanorobots to identify malignant tissues. Integrated nanosensors can be utilized for such a task in order to find intensity of E-cadherin signals. Thus they can be employed effectively for treating cancer.

Nanorobots in Cancer Detection and Treatment

Pharmacyte is a self-powered, computer controlled medical nanorobot system capable of digitally precise transport, timing, and targeted-delivery of pharmaceutical agents to specific cellular and intracellular destinations within the human body. Pharmacytes escape the phagocytic process as they will not embolize small blood vessels because the minimum viable human capillary that allows passage of intact erythrocytes and white cells is 3–4 micronmeter in diameter, which is larger than the largest proposed Pharmacyte. Pharmacytes will have many applications in nanomedicine such as initiation of apoptosis in cancer cells and direct control of cell signaling processes.

Pharmacytes could also tag target cells with biochemical natural defensive or scavenging systems, a strategy called "phagocytic flagging". For example, novel recognition molecules are expressed on the surface of apoptotic cells. In the case of T lymphocytes, one such molecule is phosphatidylserine, a lipid that is normally restricted to the inner side of the plasma membrane [1m] but, after the induction of apoptosis, appears on the outside.

Cells bearing this molecule on their surface can then be recognized and removed by phagocytic cells. Seeding the outer wall of a target cell with phosphatidylserine or other molecules with similar action could activate phagocytic behaviour by macrophages, which had mistakenly identified the target cell as apoptotic substances capable of triggering a reaction by the body Pharmacytes would be capable of carrying up to approximately 1cubicmeter of pharmaceutical payload stored in onboard tanks that are mechanically offloaded using molecular sorting pumps operated under the control of an onboard computer.

Depending on mission requirements, the payload can be discharged into the proximate extracellular fluid or delivered directly into the cytosol using a transmembrane injector mechanism. If needed for a particular application, deployable mechanical cilia and other locomotive systems can be added to the Pharmacyte to permit transvascular and transcellular mobility, thus allowing delivery of pharmaceutical molecules to specific cellular and even intracellular addresses with negligible error.

Pharmacytes, once depleted of their payloads or having completed their mission, would be recovered from the patient by conventional excretory pathways. The nanorobots might then be recharged, reprogrammed and recycled for use in a second patient who may need a different pharmaceutical agent targeted to different tissues or cells than in the first patient.

Nanorobots in the Diagnosis and Treatment of Diabetes

Glucose carried through the blood stream is important to maintain the human metabolism working healthfully, and its correct level is a key issue in the diagnosis and treatment of diabetes. Intrinsically related to the glucose molecules, the protein hSGLT3 has an important influence in maintaining proper gastrointestinal cholinergic nerve and skeletal muscle function activities, regulating extracellular glucose concentration. The hSGLT3 molecule can serve to define the glucose levels for diabetes patients.

The most interesting aspect of this protein is the fact that it serves as a sensor to identify glucose. The simulated nanorobot prototype model has embedded Complementary Metal Oxide semi-conductor (CMOS) nanobioelectronics.

It features a size of ~2 micronmeter, which permits it to operate freely inside the body. Whether the nanorobot is invisible or visible for the immune reactions, it has no interference for detecting glucose levels in blood stream. Even with the immune system reaction inside the body, the nanorobot is not attacked by the white blood cells due biocompatibility.

For the glucose monitoring the nanorobot uses embedded chemosensor that involves the modulation of hSGLT3 protein glucosensor activity. Through its onboard chemical sensor, the nanorobot can thus effectively determine if the patient needs to inject insulin or take any further action, such as any medication clinically prescribed. The image of the NCD simulator workspace shows the inside view of a venule blood vessel with grid texture, red blood cells (RBCs) and nanorobots. They flow with the RBCs through the bloodstream detecting the glucose levels.

At a typical glucose concentration, the nanorobots try to keep the glucose levels ranging around 130 mg/dl as a target for the Blood Glucose Levels (BGLs). A variation of 30mg/dl was adopted as a displacement range, though this can be changed based on medical prescriptions. In the medical nanorobot architecture, the significant measured data can be then transferred automatically through the RF signals to the mobile phone carried by the patient. At any time, if the glucose achieves critical levels, the nanorobot emits an alarm through the mobile phone.

Controlling Glucose Level using Nanorobots

In the simulation, the nanorobot is programmed also to emit a signal based on specified lunch times, and to measure the glucose levels in desired intervals of time. The nanorobot can be programmed to activate sensors and measure regularly the BGLs early in the morning, before the expected breakfast time. Levels are measured again each 2 hours after the planned lunchtime. The same procedures can be programmed for other meals through the day times.

A multiplicity of blood borne nanorobots will allow glucose monitoring not just at a single site but also in many different locations simultaneously throughout the body, thus permitting the physician to assemble a whole-body map of serum glucose concentrations.

Examination of time series data from many locations allows precise measurement of the rate of change of glucose concentration in the blood that is passing through specific organs, tissues, capillary beds, and specific vessels. This will have diagnostic utility in detecting anomalous glucose uptake rates which may assist in determining which tissues may have suffered diabetes-related damage, and to what extent.

Other onboard sensors can measure and report diagnostically relevant observations such as patient blood pressure, early signs of tissue gangrene, or changes in local metabolism that might be associated with early-stage cancer. Whole-body time series data collected during various patient activities levels could have additional diagnostic value in assessing the course and extent of disease.

This important data may help doctors and specialists to supervise and improve the patient medication and daily diet. This process using nanorobots may be more convenient and safe for making feasible an automatic system for data collection and patient monitoring. It may also avoid eventually infections due the daily small cuts to collect blood samples, possibly loss of data, and even avoid patients in a busy week to forget doing some of their glucose sampling.

These Recent developments on nanobioelectronics show how to integrate system devices and cellular phones to achieve a better control of glucose levels for patients with diabetes.

Respirocyte

The artificial mechanical red cell, "Respirocyte" is an imaginary nanorobot, floats along in the blood stream. These atoms are mostly carbon atoms arranged as diamond in a porous lattice structure inside the spherical shell. The Respirocyte is essentially a tiny pressure tank that can be pumped full of oxygen (O_2) and carbon dioxide (CO_2) molecules.

Later on, these gases can be released from the tiny tank in a controlled manner. The gases are stored onboard at pressures up to about 1000 atmospheres. Respirocyte can be rendered completely nonflammable by constructing the device internally of sapphire, a flameproof material with chemical and mechanical properties otherwise similar to diamond.

There are also gas concentration sensors on the outside of each device. When the nanorobot passes through the lung capillaries, O_2 partial pressure is high and CO_2 partial pressure is low, so the onboard computer tells the sorting rotors to load the tanks with oxygen and to dump the CO_2. When the device later finds itself in the oxygen-starved peripheral tissues, the sensor readings are reversed.

That is, CO_2 partial pressure is relatively high and O_2 partial pressure relatively low, so the onboard computer commands the sorting rotors to release O_2 and to absorb CO_2.Respirocytes mimic the action of the natural hemoglobin-filled red blood cells. But a Respirocyte can deliver 236 times more oxygen per unit volume than a natural red cell.

This nanorobot is far more efficient than biology, mainly because its diamondoid construction permits a much higher operating pressure. So the injection of a 5 cm^3 dose of 50% Respirocyte aqueous suspension into the bloodstream can exactly replace the entire O_2 and CO_2 carrying capacity of the patient's entire 5,400 cm^3 of blood. Respirocyte will have pressure sensors to receive acoustic signals from the doctor, who will use an ultrasound-like transmitter device to give the Respirocyte commands to modify their behaviour while they are still inside the patient's body.

Artificial Phagocytes

A microbivore has been described, whose primary function is to destroy microbiological pathogens found in the human bloodstream, using the "digest and discharge" protocol. Nanorobotic artificial hypothetical phagocytes called "microbivores" could patrol the bloodstream, seeking out and digesting unwanted pathogens including bacteria, viruses, or fungi. Microbivores when given intravenously (I.V) would achieve complete clearance of even the most severe septicemic infections in hours or less.

This is far better than the weeks or months needed for antibiotic-assisted natural phagocytic defences. The nanorobots do not increase the risk of sepsis or septic shock because the pathogens are completely digested into harmless simple sugars, monoresidue amino acids, mononucleotides, free fatty acids and glycerol, which are the biologically inactive effluents from the nanorobot.

A Hypothetical Mobile Cell-Repair Nanorobot

Another nanorobot, the Chromallocyte would replace entire chromosomes in individual cells thus reversing the effects of genetic disease and other accumulated damage to our genes, preventing aging.

Chromallocyte is a hypothetical mobile cell-repair nanorobot capable of limited vascular surface travel into the capillary bed of the targeted tissue or organ, followed by extravasation, histonatation, cytopenetration, and complete chromatin replacement in the nucleus of one target cell, and ending with a return to the bloodstream and subsequent extraction of the device from the body, completing the cell repair mission."

Inside a cell, a repair machine will first size up the situation by examining the cell's contents and activity, and then take action. By working along molecule-by-molecule and structure-by-structure, repair machines will be able to repair whole cells. By working along cell-by-cell and tissue-by-tissue, they (aided by larger devices, where need be) will be able to repair whole organs. By working through a person organ by organ, they will restore health. Because molecular machines will be able to build molecules and cells from scratch, they will be able to repair even cells damaged to the point of complete inactivity.

Further Applications of Nanorobots

Nanorobots could be used to maintain tissue oxygenation in the absence of respiration, repair and recondition the human vascular tree eliminating heart disease and stroke damage, perform complex nanosurgery on individual cells, and instantly staunch bleeding after traumatic injury. Monitoring nutrient concentrations in the human body is a possible application of nanorobots in medicine. One of interesting nanorobot utilization is also to assist inflammatory cells (or white cells) in leaving blood vessels to repair injured tissues. Nanorobots might be used as well to seek and break kidney stones. Nanorobots could also be used to process specific chemical reactions in the human body as ancillary devices for injured organs. Nanorobots equipped with nanosensors could be developed to deliver anti-HIV drugs.

Another important capability of medical nanorobots will be the ability to locate stenosed blood vessels, particularly in the coronary circulation, and treat them mechanically, chemically, or pharmacologically.

To cure skin diseases, a cream containing nanorobots may be used. It could remove the right amount of dead skin, remove excess oils, add missing oils, apply the right amounts of natural moisturizing compounds, and even achieve the elusive goal of 'deep pore cleaning' by actually reaching down into pores and cleaning them out. The cream could be a smart material with smooth-on, peel-off convenience. A mouthwash full of smart nanomachines could identify and destroy pathogenic bacteria while allowing the harmless flora of the mouth to flourish in a healthy ecosystem. Further, the devices would identify particles of food, plaque, or tartar, and lift them from teeth to be rinsed away. Being suspended in liquid and able to swim about, devices would be able to reach surfaces beyond reach of toothbrush bristles or the fibres of floss.

As short-lifetime medical nanodevices, they could be built to last only a few minutes in the body before falling apart into materials of the sort found in foods. Medical nanodevices could augment the immune system by finding and disabling unwanted bacteria and viruses. When an invader is identified, it can be punctured, letting its contents spill out and ending its effectiveness. If the contents were known to be hazardous by themselves, then the immune machine could hold on to it long enough to dismantle it more completely.

Devices working in the bloodstream could nibble away at arteriosclerotic deposits, widening the affected blood vessels. Cell herding devices could restore artery walls and artery linings to health, by ensuring that the right cells and supporting structures are in the right places.

This would prevent most heart attacks. Nanorobots could be used in precision treatment and cell targeted delivery, in performing nanosurgery, and in treatments for hypoxemia and respiratory illness, dentistry, bacteremic infections, physical trauma, gene therapy via chromosome replacement therapy and even biological aging. It has been suggested that a fleet of nanorobots might serve as antibodies or antiviral agents in patients with compromised immune systems, or in diseases that do not respond to more conventional measures. There are numerous other potential medical applications, including repair of damaged tissue, unblocking of arteries affected by plaques, and perhaps the construction of complete replacement body organs.

Nanoscale systems can also operate much faster than their larger counterparts because displacements are smaller; this allows mechanical and electrical events to occur in less time at a given speed. Nanotechnology as a diagnostic and treatment tool for patients with cancer and diabetes showed how actual developments in new manufacturing technologies are enabling innovative works which may help in constructing and employing nanorobots most effectively for biomedical problems.

Nanorobots applied to medicine hold a wealth of promise from eradicating disease to reversing the aging process (wrinkles, loss of bone mass and age-related conditions are all treatable at the cellular level); nanorobots are also candidates for industrial applications. The advent of molecular nanotechnology will again expand enormously the effectiveness, comfort and speed of future medical treatments while at the same time significantly reducing their risk, cost, and invasiveness.

CELLULAR BIOSCANNING

The goal of cellular bioscanning is the noninvasive and non-destructive in vivo examination of interior biological structures. One of the most common nanomedical sensor tasks is the scanning of cellular and subcellular structures. Such tasks may include localization and examination of cytoplasmic and nuclear membranes, as well as the identification and diagnostic measurement of cellular contents including organelles and other natural molecular devices, cytoskeletal structures, biochemical composition and the kinetics of the cytoplasm.

The precision and speed of medical nanodevices is so great that they can provide a surfeit of detailed diagnostic information well beyond that which is normally needed in classical medicine for a complete analysis of somatic status.

Except in the most subtle cases a malfunction in any one component of the cellular machinery normally provokes a cascade of pathological observables in many other subsystems. Detection of any one of these cascade observables, if sufficiently unique and well-defined, may provide adequate diagnostic information to plan a proper reparative procedure.

DNA is one of the few cellular components that is regularly inspected and repaired. Most homeostatic systems adopt a more simple philosophy of periodic replacement of components regardless of functionality. Nanomedicine allows the philosophy of inspection and repair to be extended to all cellular components. The following discussion briefly describes a few of the sensory techniques that may be useful in achieving this objective.

CELLULAR TOPOGRAPHICS

Tactile topographic scanning provides the most direct means for examining cellular structures in vivo. For example, ex vivo live cell scanning in air by commercially available atomic force microscopy using a 20–40 nm radius tip allows nondestructive feature resolutions of ~50 nm across the top 10 nm of the cell.

Investigators have used the scanning tip to punch a hole in the cell, then pull back and scan the breach, observing the membrane heal itself via self-assembly in real time. Up to ~KHz scanning frequencies are possible, with up to 1024 data points per scan line. Assuming a scan rate of ~10^6 pixels/sec, a micron-scale nanodevice, once securely anchored to the surface of a

(~20 micron)^3 human cell, could employ a tactile scanning probe to image the 0.1% of plasma membrane lying within its (1.4 micron)^2 vicinity in ~2 sec to ~1 nm^2 resolution (~1 mm/sec tip velocity), or ~50 sec to ~0.2 nm (*e.g.*, atomic) resolution (~0.2 mm/sec tip velocity).

Inside the cell, and again post-anchoring, an entire 6 micron^2 mitochondrial surface could be imaged to atomic resolution in ~100 sec; the surface of a 100-nm length of 25-nm diameter microtubule could be atomically resolved in 0.2 sec—though of course in a living cell these structures may be changing dynamically during the scanning process.

From Eqn., continuous power dissipation of a (0.1 micron)^2 scan head moving at 1 mm/sec through water is ~0.002 pW (~kT/pixel at ~10^6 pixels/sec), though of course the energy cost of sensing, recording, and processing each pixel must be at least ~10 kT/pixel (ignoring the possibility of pre-computational image compression), so the total scanner power draw could be as high as 0.02–0.1 pW in continuous operation. Special scanning tips and techniques should allow topographic, roughness, elastic, adhesive, chemical, electrostatic, conductance, capacitance, magnetic, or thermal surface properties to be measured.

For large (~0.1–1 micron) cellular components, identification and preliminary diagnosis of improper structure may be possible using measured surface characteristics which may be matched to entries in an extensive onboard library, perhaps combined with dynamic monitoring of anomalies in cross-membrane molecular traffic using chemical nanosensors.

Most membranes are self-sealing, so it should also be possible to gently insert a telescoping member into the target organelle, which member then slowly reticulates and extrudes smaller probes with sensory tips in a fixed pattern and step size, allowing the acquisition of detailed internal structural and compositional information. Small (~1–10 nm) protein-based components of the cell such as enzymes, MHC carriers, and ribosomes are self-assembling or require only modest assistance to self-assemble. Reversible mechanical denaturation of these smaller proteins, using diamondoid probe structure to displace water molecules and by using specialized handling tools akin to functionalized AFM tips and molecular clamps, may be followed by precise nondestructive amino acid sequencing to allow identification and diagnostic compositional analysis.

Afterward, the protein molecule may be refolded back into its original minimum-energy conformation, possibly with the assistance of chaperone-like structures in some cases.

Acoustic Microscopy

Acoustic microscopy is another noninvasive scanning technique that can provide nanomedically useful spatial resolutions. Frequencies are in the GHz range, far exceeding the relaxation time of the protoplasm. A cryogenic acoustic microscope operated at 8 GHz has demonstrated 20 nm lateral resolution in liquid helium. By 1998, the best resolution achieved with water as the coupling fluid has been 240 nm at a frequency of 4.4 GHz. Operated in water at 310 K, a nanomechanical 1.5 GHz acoustic microscope would achieve a far-field minimum lateral resolution l/2 ~ 500 nm, ~10^5 voxels per human cell, sufficient to locate and count all major organelles and a few intermediate-scale structures.

Scanning acoustic microscopes (SAM) operated at 2 GHz in reflection mode (which allows detection of interference effects) achieve 30–50 nm resolution in the direction of the acoustical axis, and the Subtraction SAM approach reveals topographical deviations of 7.5 nm at 1 GHz.

It has been proposed that picosecond ultrasonics could be used to obtain an image of the cytoskeleton with detail comparable to that of conventional X-ray images of a human skeleton. Power requirements are a significant constraint on acoustic reflection microscopy (echolocation).

However, cells and body tissues are mesoscopic "junkyards"—highly heterogeneous media which may produce large numbers of nontarget scattering events, thus increasing the difficulty of extracting signal from noise.

Additional complications arise due to:

- Scattering on rough surfaces;

- Rapid pressure variations with range in the Fresnel zone or near field of the transmitted signal;
- Cytoplasmic viscosity inhomogeneities due to the asymmetric arrangement of cytoskeletal structure, granules, vacuoles, and the endomembrane system; and
- Variation in cytoplasmic Young's modulus due to time-varying tensions in semirandomly-distributed cytosolic fibrillar elements which alter elasticity and thus the local speed of sound.

Magnetic Resonance Cytotomography

Electrostatic scanning is largely ineffective because of Debye-Huckel shielding. Magnetic stray field probes allow resolution of 10 nm physical features, but only for materials with substantial magnetic domains—which most biological substances lack. But nuclear magnetic resonance (NMR) imaging may allow cellular tomography by creating 3-D proton (hydrogen atom) density maps. Atomic density maps of other biologically important elements with nonzero nuclear magnetic moments (including D, Li^6, B^{10}, B^{11}, C^{13}, N^{14}, N^{15}, O^{17}, F^{19}, Na^{23}, Mg^{25}, P^{31}, Cl^{35}, K^{39}, Fe^{57}, Cu^{63} and Cu^{65}) may also be compiled. For example, sodium imaging is already used clinically to assess brain damage in patients with strokes, epilepsy, and tumors. In a hypothetical NMR cytotomographic nanoinstrument, a large permanent magnet is positioned near the surface of the cell or organelle to be examined. This creates a large static background magnetic field that polarizes the protons. The spatial gradient of this field establishes a unique resonant frequency, called the Larmor frequency, within each isomagnetic surface throughout the test volume.

A second time-varying magnetic driver field is then scanned through the full range of resonant frequencies, exciting into resonance the protons in each isomagnetic surface, in turn. Depending on the sensor implementation chosen, each resonance detected may cause absorption of driver field energy, an increase in measured impedance, or even a return echo of magnetic energy as the excited protons relax to equilibrium in ~1 sec. The large polarizing magnet is then rotated to a new orientation, moving the isomagnetic surfaces to new positions within the test volume, and the scan is repeated. A 3-D map of the spatial proton distribution may be computed after several scan cycles.

Near-Field Optical Nanoimaging

Electromagnetic waves of optical wavelength l that interact with an object are diffracted into two components, called "far-field" and "near-field." The propagation of electromagnetic radiation over distances z > l acts as a spatial filter of finite bandwidth, resulting in the familiar diffraction-limited resolution ~λ/2.

Classical optics is concerned with this far-field regime with low spatial frequencies < 2/l, and conventional optical imaging will be difficult at the cellular level in vivo. (Short-wave X-rays will damage living biological cells.) Information about the high spatial frequency components of the diffracted waves is lost in the far-field regime, so information about sub-wavelength features of the object cannot be retrieved in classical microscopy.

However, for propagation over distances z << λ, far higher spatial frequencies can be detected because their amplitudes are then of the same order as the sample (z = 0). This second diffraction component is the "near-field" evanescent waves with high spatial frequencies >2/λ. Evanescent waves are confined to subwavelength distances from the object. Thus a localized optical probe, such as a subwavelength aperture in an opaque screen, can be scanned raster fashion in this regime to generate an image with a resolution on the order of the probe size.

The original Near-field Scanning Optical Microscope (NSOM) surpassed the classical diffraction limit by operating an optical probe at close proximity to the object. The NSOM probe uses an aluminum-coated "light funnel" scanned over the sample. Visible light emanates from the narrow end (~20 nm in diameter) of the light funnel and either reflects off the sample or travels through the sample into a detector, producing a visible light image of the surface with ~12 nm resolution at $\lambda = 514.5$ nm provided the distance between light source and sample is very short, about 5 nm, with signal intensities up to 10^{11} photons/sec (~50 nanowatts).

This represents a resolution of ~$\lambda/40$. The near-field acoustic equivalent is found in the medical stethoscope, which exhibits a resolution of ~$\lambda/100$.Applications include dynamical studies at video scanning rates, low-noise high-resolution spectroscopy, and differential absorption measurements.

Optical imaging of individual dye molecules has already been demonstrated, with the ability to determine the orientation and depth of each target molecule located within ~30 nm of the scanned surface. Molecules with nanometer-scale packing densities have been resolved to ~0.4-nm diameters using STMs to create photon emission maps, and laser interferometric NSOMs have produced clear optical images of dispersed oil drops on mica to ~1 nm resolution.

Live specimen 250-nm optical sectioning for three-dimensional dynamic imaging has also been demonstrated. Submicron laser emitters have been available since the late 1980s. It should be possible to use NSOM-like nanoprobes to optically scan the surfaces of cells or organelles to <~1 nm resolution, mapping their topography and spectroscopic characteristics to depths of tens of nanometers without penetrating the surface.

However, a proper membrane-sealing invasive light funnel ~20 nm in diameter might not seriously disrupt some cellular or organelle membranes and thus could be inserted into the interiors of these bodies or through the cytoskeletal interstices to permit deeper volumetric scanning.

The thermal conductivity of water at 310 K is 0.623 watts/m-K and the energy per 500-nm photon is 400 zJ, so for 1 micron^3 of watery tissue the maximum scan rate is ~35 micron^3/sec-K, if e^{SNR} photons are used to image each 1 nm^3 voxel with SNR = 2. Thus a 1-sec volumetric optical scan of an aqueous ~1-micron^3 sample volume to 1 nm^3 resolution requires a ~1 nanowatt scanner running at ~GHz bit rates, raising sample volume temperature by ~0.03 K.

NSOM permits the determination of five of the six degrees of freedom for each molecule, lacking only the optically inactive rotation around the dipole axis. In principle, it should be possible to produce <~1 nm resolution near-field optical scans of in situ protein molecules, since with atomically precise fabrication and single lines of atoms as conductors a minimum light guide (metal-dielectric-metal) is 3 atoms wide.

Given a molecular laser and adequate collimation, photons passing through folded proteins will scatter according to the molecular structure. Detection of sufficient photons comprising these scattering patterns should allow the noninvasive determination of protein structure; polarized photons provide information on chirality. Absorption and fluorescent signals will be visible from phenyl rings, tryptophan, and bound cofactors such as ATP and adenine (which is fluorescent). Positions of monoclonal antibodies on virus surfaces are now identifiable experimentally using NSOM; it should be possible to map binding sites on virus and cell surfaces using fluorescently labelled antibodies.

Single molecule detection has been proposed as a tool for rapid base-sequencing of DNA. Other optically-based cellular imaging techniques must be distinguished. Optical Coherence Tomography (OCT) uses Michelson interferometry to achieve ~10 micron spatial resolutions over tissue depths of 2-3 mm in nontransparent tissue in the near infrared.

However, OCT requires numerous physical components not easily implemented on a micron-size detector, femtosecond pulse shaping, and illumination power levels of ~10^7 watts/m^2 >> 100 watts/m^2 "safe" continuous limit in tissue. Bioluminescence techniques in which the light source is placed inside the tissue has a spatial resolution limited to ~10% of the depth, or ~10 microns resolution at a ~100 micron depth.

Coherent anti-Stokes Raman scattering (CARS) already permits organelle imaging in living cells and can in theory create a point-by-point chemical map of a cell using two intersecting lasers. Three-dimensional observation of microscopic biological nonliving structures by means of X-ray holography requires a high degree of spatial coherence and good contrast between target and surroundings. Good contrast may be achieved in the wavelength range between the K absorption edges of carbon (l = 4.37 nm) and oxygen (l =2.33 nm), the "water window" where carbon-containing biological objects absorb radiation efficiently but water is relatively transparent. A 50-micron diameter emitter has been tested that uses near-IR 5-femtosecond laser pulses impinging upon a helium gas target to create a well-collimated (<1 milliradian) beam of coherent soft X-rays at a 1 KHz repetition rate producing a brightness of 5×10^8 photons/mm^2-milliradian^2-sec, a peak X-ray intensity of $>10^{10}$ watts/m^2 on the propagation axis behind the He target.

Cell Volume Sensing

Many cellular parameters need not be measured directly in order to be detected by a medical nanodevice. Cell volume sensing is a case in point. An intracellular nanodevice can indirectly monitor changes in the volume of the cell in which it resides by one of two methods.

First, measurements of the mechanical deformation of the cellular membrane, stretch-activated channels, or cytoskeletal strains and structural changes are quite sensitive to alterations in total cell volume.

Second, concentration or dilution of the cytoplasmic environment through cell shrinkage or swelling leads to the activation of various volume-regulation responses which may be detected by the nanorobot. Changes in the concentration of soluble cytosolic proteins may nonspecifically affect enzyme activity via "macro- molecular crowding". Minor changes in cell volume can cause severalfold changes in ion transport.

Cellular signalling entities that have been linked to the transduction and amplification of the primary volume signal include Ca^{++} transients, phosphoinositide turnover, eicosanoid metabolism, kinase/phosphatase systems such as JNK and p38, cAMP, and G-proteins. These signal amplification pathways may be monitored using chemical concentration sensors aboard the medical nanodevice, allowing the nanorobot to eavesdrop on the natural sensory channel traffic of the cell.

NONINVASIVE NEUROELECTRIC MONITORING

Noninvasive measurement of axonal traffic within nerve bundles will require multiple sensors and greater sensitivity to compensate for shielding by the perineurium, a tight resistive sheath enclosing the bundle ~20 microns thick with resistivity ~4000 ohm-cm.

Electric Field Neurosensing

The "typical" ~20 micron human neuron discharges 5-100 times per second, moving from -60 mV potential to +30 mV potential in ~10^{-3} sec. Thus the variation in electric field at the axonal surface is ± 4500 volts/m. Since electric field sensors can detect 100 volt/m fields up to GHz frequencies, an electric sensor attached by circumaxonal cuff or pressed against the axonal surface (possibly at the node of Ranvier) should readily detect each action potential discharge. (The diameter of human nerve axons is 0.1-20 microns.)

By 1998 silicon-to-neuron extracellular junctions already permitted direct stimulation of individual nerve cells in vitro without killing the cells, and extracellular electrodes were commonly used to detect neuronal electrical activity noninvasively, both in vivo and in vitro.

Artificial electric fields may also be employed to trigger or moderate neural signals. Cell membrane capacitance is typically ~0.01 picofarads/micron^2, and varies with the state of the health of the cell.

Magnetic Field Neurosensing

The magnetic flux density caused by a single action potential discharge is ~0.1 microtesla at the axonal surface, which may be detected by a "compass oscillator" type magnetosensor with a ~KHz maximum sampling rate. It might be possible for artificial magnetic fields to directly influence neural transmissions. Even a static 65 millitesla field has been shown to reduce frog skin Na^+ transport by 10-30%.

Each neuronal discharge develops an electrical energy of ~20 picojoule (~10^{10} kT), far smaller than the magnetic energy stored in a B = 1.4 tesla field of a permanent micromagnet traversing an L^3 = (20 micron)^3 volume which from Eqn. is $B^2 L^3/2 \mu_0$ ~ 6000 pJ. If properly manipulated, such a field may be sufficient to enhance, modulate, or extinguish a passing neural signal.

Neurothermal Sensing

While neurons come in many shapes and sizes, our "exemplar" ~14,000-micron^3 neuron discharging ~90 mV into an input impedance of ~500 Kohms produces ~0.2 microampere current per pulse and generates a continuous (average) 100-300 pW of waste heat as measured experimentally. The discharge rate of 5-100 Hz can produce brief surges up to ~2000 pW during a high-frequency train, but the duty cycle of such trains is far less than 100%, reducing time-averaged dissipation to the observed 100-300 pW range.

Single impulses are measured experimentally to produce 2-7 microkelvin temperature spikes in cold or room-temperature mammalian non-myelinated nerve fibres and ~23 microkelvins in non-myelinated garfish olfactory nerve fibres at an energy density ranging from 270-1670 joules/m^3-impulse from 0-20°C.

In non-myelinated fibres the initial heat occurs in two temperature-dependent phases: a burst of positive heat, followed by rapid heat reabsorption. The positive heat derives from the dissipation of free energy stored in the membrane capacity, and from the decrease in entropy of the membrane dielectric with depolarization.

An L ~ 20-micron neuron in good thermal contact with an aqueous heat sink at 310 K has thermal conductance $L K_t$ ~ 10^{-5} watts/K, so trains of 5-100 Hz impulses lasting 1 second should raise cellular temperature by 10-30 microkelvins; up to ~200 microkelvin thermal spikes from such trains have been observed experimentally.

These events are easily detectable by nanoscale thermal sensors capable of ~1 microkelvin sensitivity up to ~1 KHz. The ~microkelvin heat signature of individual impulses or very short pulse trains can probably be temporally resolved because the minimum pulse repetition time is ~10 millisec (at 100 Hz) which is much longer than the thermal time constant for an L ~ 20-micron neuron which is $t_{EQ} \sim L^2 C_V/K_t \sim 3$ millisec.

Direct Synaptic Monitoring

The synaptic cleft between the axonal presynaptic terminal and the dendritic postsynaptic membrane is 10-20 nm in most synapses, although in the vertebrate myoneural junction it may be as large as 100 nm. Contact area per bouton is ~1 micron^2, giving a total gap volume of ~10^7-10^8 nm^3. The density of acetylcholine receptors is highest in muscles along the crests and upper thirds of the junctional folds (~10,000/micron^2), and is lowest in the extrasynaptic regions (~5/micron^2).

Each action potential discharge triggers the release of ~10^4-10^5 molecules of acetylcholine into the gap volume of an active neuromuscular junction (diffusion time ~1 microsec), raising c_{ligand} from near zero to ~3 × 10^{-4} molecules/nm^3 (~0.0005 M) in ~1 millisec, followed by near-complete hydrolyzation by acetylcholinesterase during the 1-2 millisec refractory period.

Into the gap volume may easily be inserted a ~10^5 nm^3 neurotransmitter concentration sensor able to measure ~100 acetylcholine molecules in ~1 millisec, thus detecting pulses at the fastest discharge rate. A similar device could be used to precisely regulate neuro-transmitter concentration at the junction, and hence the neural signal itself, under nanodevice control. Simple electrochemical and mechanochemical artificial synapses have been demonstrated.

RF AND MICROWAVE OSCILLATIONS

Interestingly, this frequency span is very close to the maximum trigger/reset frequency for bioelectronic molecules. Note that (~100 mV) (1.6 x 10^{-19} coul) ~ 4 kT, so a membrane molecule with a single charge on either end should be reliably reoriented by a depolarization wave, coupling pressure waves and electrostatic field fluctuations. However, the direct detection of 10-100 GHz millimeter radiation by non-nanotechnological means is experimentally difficult and controversial because the tests must be performed in vivo in close proximity to an actively metabolizing cell in water—and water strongly absorbs microwaves over macroscale ranges. Nevertheless, active cells have shown enhanced Raman anti-Stokes scattering, an effect ascribed to the converse of the Frohlich oscillations. In one study, the normalized growth rate of yeast cultures was enhanced or inhibited when irradiated by CW microwave fields of ~30 watts/m^2 of various frequencies; growth rate data spanning 62 separate runs revealed a repeatable frequency-dependent spectral fine structure with six distinct peaks of width ~10 MHz near 42 GHz.

Investigations of related phenomena are ongoing and voluminous; the interested reader should peruse *Bioelectromagnetics*, the archival journal of this field. 100 GHz waves attenuate only ~1% after passing through ~3 microns of soft tissue. A single electron injected into an integral membrane protein could act as an oscillating dipole, making a 300 volt/m signal 10 nm from the protein antenna with an energy transfer of ~0.004 kT per cycle; ~1000 oscillating electrons could produce a measurable field. A 20-micron diameter cell modeled as a nonuniform spherical dipole layer with transmembrane dipoles located 10 nm apart and embedded in a dissipative medium could produce 10^2–10^5 volt/m microwave fields 1-10 microns from the cell surface.

Thus, a variety of rf and microwave electromagnetic emanations may in theory be detectable both within and nearby living cells which could prove diagnostic of numerous internal states.

Such states may include cytoskeletal dynamics, metabolic rates, plasmon-type excitations due to the collective motion of ions freed in chemical reactions, positional, rotational or conformational changes in biological macromolecules and membranes, internal movements of organelles and nerve traffic conduction, cellular pinocytosis, cellular reproduction events, cell membrane identity, and cell-cell interactions.

MACROSENSING

Macrosensing is the detection of global somatic states (inside the human body) and extrasomatic states (sensory data originating outside the human body). While the treatment here is necessarily incomplete, the discussion nevertheless gives a good feel for the kinds of environmental variables that internally-situated nanodevices could sense.

Not all capabilities outlined here need be available on every nanorobot, since injection of a cocktail of numerous distinct but mutually cooperative machine species allows designers to take full advantage of the benefits of functional specialization.

In many cases, a given environmental variable can be measured by several different classes of sensor device. However, since these devices are microscopic it is in theory possible to operationalize almost all of the macrosensing capabilities described below in one patient using just a billion devices (~10 mm^{-3} whole-body deployment density), a total volume of ~1 mm^3 of nanorobots or ~0.1% of the typical ~1 cm^3 therapeutic dose.

BLOOD PRESSURE

Blood pressure ranges from 0.1– 0.2 atm in the arteries to as low as 0.005 atm in the veins. The systolic/diastolic differential ranges from 0.05–0.07 atm in the aorta and 0.01–0.02 atm in the pulmonary artery, falling to 0.001–0.003 atm in the microvessels, or 0.003–0.005 atm if the precapillary sphincter is dilated.

In venous vessels, pulse fluctuations are 0.002–0.010 atm in the superior vena cava, 0.004–0.006 atm in the subclavian vein, ~0.004 atm in venules generally, and ~0.0005 atm in the brachial vein. There is also a ~0.05 Hz random fluctuation in the microvessels with amplitude on the order of 0.004–0.007 atm. Both blood pressure and pulse rate can be reliably monitored by a medical nanodevice virtually anywhere in the vascular system using a (68 nm)^3 pressure sensor with ~0.001 atm sensitivity.

Pulse propagation through body tissue is somewhat muted due to absorption in compressible fatty membranes, but most cells lie within 1–3 cell-widths of a capillary so the cardiac acoustic signal should still be measurable using more sensitive detectors. The time-averaged interstitial pressure in subcutaneous tissue is 0.001–0.004 atm. Arterial pulse waves (vascular oscillations) carry subtle messages about the health of internal organs and the arterial tree.

The idea of using pulse waves for diagnosis has a long history dating back 2000 years in China. Waves detected by manual probing are classified using such subjective and qualitative descriptors as floating, deep, hidden, rapid, slow, moderate, feeble, replete, full, thready, faint, weak, soft, slippery, hesitant, hollow, firm, long, short, swift, running, intermittent, uneven, taut, string-tight, gigantic, or tremulous.

Abnormal waves were empirically related to disease states. Wave data gathered by nanodevices could make possible a theoretically sound, quantitative system of noninvasive observation, classification, and diagnosis as a supplement to other nanomedical tools.

Respiratory Audition

The variation of mechanical pressure over a complete respiratory cycle is ~0.003 atm in the pleura, ~0.002 atm at the alveoli, detectable by nanomedical pressure sensors positioned in the vicinity of the respiratory organs. Holding a deep breath further stretches the pulmonary elastic tissue, up to 0.02 atm. However, turbulent flows at Reynolds number $N_R > 2300$ in the trachea, main bronchus and lobar bronchus produce a whooshing noise that may be the loudest noncardiac sound in the human torso during conventional auscultation.

The energy dissipation for turbulent flow in a tube is:

$$P_{turb} = P_{lam} Z = 8 \pi_{air} v^2 L Z \text{ (watts)}$$

where P_{lam} is the dissipation for laminar (Poiseuillean) flow in a long circular cylindrical tube of length L, v is the mean flow velocity, and turbulence factor $Z - 0.005 (N_R^{3/4} - (2300)^{3/4})$, a well-known empirical formula.

For $\eta_{air} = 1.83 \times 10^{-5}$ kg/m-sec for room-temperature air (20°C), $P_{turb} = 0.87$ milliwatts for the trachea; $P_{turb} = 0.66$ milliwatts for the main bronchus (L = 0.167 m, v = 4.27 m/sec and $N_R = 3210$); and $P_{turb} = 0.09$ milliwatts for the lobar bronchus (L = 0.186 m, v = 4.62 m/sec and N_R = 2390), totalling ~1.6 milliwatts acoustic emission from a ~120 cm^3 upper tracheobroncheal volume.

This is a power density of 13 watts/m^3 corresponding to a pressure of 4 × 10^{-5} atm assuming a 300 millisecond measurement window at the maximum respiration rate. The amplitude of an acoustic plane wave propagating through tissue attenuates exponentially with distance due to absorption, scattering and reflection. The amplitude is given approximately by,

$$A_x = A_D e^{-\alpha F x}$$

where A_0 is the initial wave amplitude in atm, A_x is the amplitude a distance x from the source, and α is the amplitude absorption coefficient. The function F expresses the frequency dependence of the attenuation. For pure liquids, $F = F_{liq} = \nu^2$ (Hz^2); for example, $\alpha_{liq} = 2.5 \times 10^{-14}$ sec^2/m for water at room temperature. However, for soft tissues, $F = F_{tiss} \sim \nu$ (Hz).

Mechanical Body Noises

Many other mechanical body noises should be globally audible to properly instrumented medical nanodevices. If normal chewing motions (of hard foods) release 1–10 milliwatts in a ~100 cm^3 oral volume with a ~1 sec jawstroke, power density is ~10 watts/m^3 or ~10^{-4} atm of toothcrunching noise. A stomach growl registering 45 dB (vs. 30 dB whisper, 60 dB normal conversation) at 2 meters has a source power of 160 milliwatts; released from a 10 cm^3 gastric sphincter volume gives a ~2 × 10^{-6} atm acoustic wave, detectable throughout the body.

Walking and running releases 20-100 joules/footfall for a 70 kg man; assuming the energy is absorbed within a ~1 cm thickness or within ~1 second by the sole of the foot, Eqn. implies an upward-moving planar compression wave of 0.4-2.0 atm, easily detectable by acoustically instrumented nanodevices body-wide. Hand-clapping generates 0.02-0.2 atm pulses, also easily detectable.

Lesser noises including ~30 millisec hiccups at (4–60)/min, intestinal and ureteral peristalsis, sloshing of liquid stomach contents, heart murmurs, a tap on the shoulder by a friend, nasal sniffling and swallowing, clicks from picking or drumming fingernails, crepitations, manustuprations and ejaculations, the rustling noise of clothing against the skin, flapping eyelids, anal towelling, bruits (including murmurs and thrills) due to vascular lesions, dermal impact of water while showering, copulatory noises, urethral flow turbulence during urination, transmitted vibrations from musical instruments, creaking joints, and squeaking muscles can be detected locally if not globally.

Implantation of significant interconnected in vivo diamondoid structures may produce increased sensitivity to internal noises, due to the extremely low acoustic absorption coefficient of diamond.

Vocalizations

Average source power for conversational speech in air is ~10 microwatts at the vocal cords (60 dB), up to ~1000 microwatts for shouting (90 dB) and as little as 0.1 microwatts (30 dB) for whispering. Vocal cord surface area ~1 cm^2, giving an acoustic intensity I ~ 0.001–10 watts/m^2. (Using the decibel notation, dB = 10 $\log_{10}$ (I/I_0), where $I_0 \sim 5 \times 10^{-13}$ watts/m^2 in air, $I_0 \sim 1 \times 10^{-16}$ watts/m^2 in water.) In a planar traveling wave, pressure amplitude A_p (N/m^2) is related to power intensity I by,

$$A_p = (2\rho\nu_{sound} I)^{1/2} \quad (N/m^2)$$

For water at 310 K, ρ = 993.4 kg/m^3 and v_{sound} = 1500 m/sec, therefore A_p = 0.0005–0.05 atm for speech, detectable by nanodevices throughout the body due to minimal attenuation at audible frequencies. Other easily detectable vocalizations include whistling, humming, coughing, sneezing, rales, wheezing, expectorating, eructations, flatus, vomiting, hawking and noseblowing.

KINESTHETIC MACROSENSING

Using a navigational transponder network, a population of ~10^{11} nanodevices each spaced an average ~100 microns apart throughout the body tissues can determine relative location to a positional accuracy of ~3 microns and an angular accuracy of ~2 milliradian, with the data updatable once every millisecond.

Thus dispersed, the network can continuously monitor and record the relative positions of all limbs with worst-case (cumulative error) ~0.8 mm accuracy over a 2-meter span. These devices can prepare high-resolution dynamic maps of body position, velocity, acceleration and rotation, allowing precise real-time digital kinesthesia during sports or artistic activities such as gymnastics, pole vaulting, or ballet; during transportative activities such as driving cars around hairpin turns, roller-coaster rides, military aircraft maneuvers and space launches; during precision tool-using such as needle-threading, antique watch repair, or while using the fingers as measurement calipers; during self-defence activities requiring complex motions such as karate or judo; and during emergency situations such as automobile crashes and tumbling motions during falls from great heights.

All such sensory data is readily outmessaged to the patient or user in real time. Nanorobots could assist in path integration or the reconstruction of experienced limb trajectories. Given the ability to determine rotation rates and to measure applied forces and torques, the network should also be able to forecast the anticipated future positions of limbs.

By comparing these projections to actual results, the network can then infer the viscosity of the medium in which the activity is taking place, whether the environment is stable or is translating or rotating in some direction, and whether the human user is physically supported or in free-fall. Internal nanodevices can directly measure if a body is sitting, standing, laying prone, inverted, falling, floating or diving, and then communicate that information directly to the human user. Detection of patient activity states—*e.g.*, the patient is sitting, standing, or walking—can be used to control nanorobot behaviours, activate or deactivate outmessaging displays, and so forth. A simple macroscale wearable tactile "compass belt" that would convey desired geographical directions has already been proposed.

Orientational Macrosensing

Since nanopendular sensors can reliably determine the direction of the local gravity field vector to within ~2 milliradians in ~10^{-4} sec, the navigational network can poll its members and arrive at an accurate consensus on which direction is up. Nanodevices affixed to relatively stable hard body parts will exhibit more consistent orientational readings. Measurement of the gravity vector allows the vertical orientation of the rest of the body to be precisely fixed in space, especially useful for gymnasts, trapeze artists, underwater divers in murky lakes, and vestibular-impaired individuals, to whom this information may be outmessaged directly

The gravitational vector may also be indirectly measured, albeit more slowly and less accurately, by monitoring the body's natural reactions to changes in the axis of gravitational loading. For instance, the variation in head-to-toe hydrostatic pressure for a 1.7 m tall, 70 kg human standing in a 1-g gravity field is ~0.17 atm; the ability to measure a systematic cross-body differential of 10^{-6} atm allows the detection of a ~10^{-5} g change along the lengthwise aspect, or ~10^{-4} g change along the transverse aspect. Thus a continuously updated whole-body barostatic map allows the human body to serve as a three-dimensional gravity/orientation sensor.

Body Weight Measurement

A network of nanodevices can inventory the volume and density of each of the body's ~10^8 $(1\ mm)^3$ voxels using a combination of acoustic ranging, somatic mapping, and flowmetry. Each voxel is identified as fat, muscle tissue, bone mass, interstitial fluid, and so forth. These measurements allow the body's weight to be precisely computed as volume times density, rather than the usual method of determining weight using gravitational force data.

The existence of a natural physiological "ponderostat," a crudely analogous humoural body-mass detector, has been proposed; indeed, the insulin-leptin system has been found to serve a related purpose, and is easily eavesdropped by medical nanorobots.

Gravitational Geographic Macrosensing

Medical nanodevices can measure variations in the gravity field to ~10^{-6} g's for L = 20 micron gravimeters in a measurement time t_{meas} = 2–9 millisec. This implies that in vivo nanodevices can take precise measurements of their latitude and altitude relative to sea level ~100 times every second.

For comparison, in 1998 high-quality commercial gravity gradiometers measured gradients of ~10^{-9} g/meter and allowed the compilation of micro-g (~1 milligal) resolution aerial gravity maps; atom interferometers measured the gravitational acceleration of atoms to a precision of 10^{-10}.

To achieve such phenomenal positional accuracies, the nanodevice must be able to computationally resolve several complicating factors. First, localized mass concentrations representing nonuniformities in crustal density produce residuals of up to ±0.0006 m/sec^2, which may be removed from the data using a standard map of known terrestrial isostatic variations and anomalies. Indeed, matching observations to such a map could provide useful longitudinal information as well. Another complication is the variation in gravity due to tidal forces amounting to ~3×10^{-7} g's, twice daily, which lies at the limits of detectability for a 20-micron gravimeter.

Other minor geodesic and terrain-related corrections, too complicated for discussion here, may also need to be applied in certain circumstances. Note that the presence of nearby heavy objects does not influence measurement accuracy: a 100-ton building 10 meters away adds a lateral acceleration of only 7×10^{-9} g's to a human body. One final complication is that patient movements create kinematic accelerations that must be distinguished from the gravitational accelerations. Gravity readings can be corrected by taking derivatives of the signals from kinesthetic monitoring to give gravity in the reference frame of the patient's room, and many individual measurements may be averaged to improve accuracy since the gravity vector normally changes only very slowly.

VASCULAR-INTERSTITIAL

In vivo studies of the electric properties of blood vessels

shows that the walls of veins and arteries present a specific electric resistance ~200 times greater than the conductive medium of blood (plasma)—roughly 200 ohm-m vs. 0.7 ohm-m. Thus blood vessels are properly regarded as relatively insulated conducting cables which can electrically connect an injured tissue with surrounding noninjured tissue. Capillaries form an electric junction with the interstitial fluid, so the electric gradient can be canceled by ionic transport.

Cell membranes are insulating dielectrics with resistance and capacitance, penetrated by ionic channels or gates for ionic transport. Additionally, in 1941 Szent-Gyorgyi suggested the possibility of semi-conduction in proteins, a theory which has since been elaborated by many investigators.

B. Nordenstrom has proposed that the above schema represents a system for selective electrogenous mass transport of material between blood and tissue which he calls Vascular-Interstitial Closed (electric) Circuits (VICCs)—in effect, an additional electro-circulatory system operating in parallel with the well-known diffusive, osmotic, and hydrodynamic mechanisms of regular bloodstream materials transport.

As long as ions can leak through open pores and migrate through ion channels, long distance transport cannot take place and the VICC system remains primarily a local (*e.g.*, within tissues), not global, electrical circuit. Connections also exist with conductive media including cerebrospinal fluid, bile, and urine. Nanorobots capable of monitoring the status of local VICCs may rapidly and efficiently acquire a wealth of systemic information without the need for direct inspection of the affected tissues.

For example, when a working muscle produces catabolic products such as lactic acid, an electrochemical potential gradient develops between the muscle and surrounding tissue in a known manner. Injured tissues become polarized in relation to surrounding tissue, as initial catabolic degradation products acidify the tissue. Malignant neoplasms, benign neoplasms, and internally necrotic granulomas all may polarize electrically in relation to surrounding tissue; a vascular thrombus also contains ionized material. Large deflections in the diffusion potential of blood occur if blood is deoxygenated or is infected with Gram-negative bacteria.

Blood changes its electric potential from +500 mV to +1000 mV during spontaneous coagulation, and the spontaneous electric potential at the site of a crush injury of the iliac crest in rats oscillated several times between +190 mV and –60 mV over a four day experimental trial. Nordenstrom proposes that in vivo electrophoresis via the VICC system may also play a role in mediating leukotaxis. When tissue is artificially polarized by electrodes, accumulation of granulocytes with margination and development of pseudopods is the result.

In one experiment, exposure of a mesenterial membrane to a 1 microampere 1 volt field for 30 minutes stimulated diapedetic bleeding near the anode. At higher power, the arterio-capillaries contracted, narrowed, and emptied of blood cells, while the veno-capillaries and venules widened and filled with granulocytes. Monitoring such actions of local VICCs experiencing natural voltage fluctuations could alert medical nanorobots to the presence of local injuries, tissue changes, or pathologies that otherwise might go unnoticed for a considerable time.

Electric/Magnetic Geographic Macrosensing

Detection of electric fields to ~100 volts/m, the normal vertical atmospheric gradient, in theory might allow determination of aboveground altitude to ~1 m accuracy if the time between recalibrations is very brief. However, a human body which is a good conductor, standing on level ground, acquires a slight negative surface charge and becomes "part of the ground" electrically. This distorts upward the equipotentials that usually run parallel to the ground, thoroughly corrupting an altitude measurement. Nearby lightning hits and storms are detectable, which is useful.

However, medical nanodevices with appropriate magnetometers can measure variations in magnetic field to ~0.1 microtesla using a (660 nm)^3 permanent magnet sensor, in a measurement time t_{meas} ~ 0.6 millisec. Geomagnetic field maps of the Earth's surface are highly nonisotropic in both latitudinal and longitudinal directions; knowledge of the field plus the absolute direction of true north gives longitude information from the separation of the planetary magnetic and spin poles.

This implies that properly equipped in vivo nanodevices can establish their latitude and longitude on Earth's surface ~1000 times every second by this means. In particular, the horizontal component of the geomagnetic field vector ranges from 0–41 microtesla from magnetic pole to magnetic equator; the independent vertical field component ranges from 0–70 microtesla. Thus a 0.1-microtesla sensor should resolve each component to an accuracy of ~11 arcmin or ~20 km in both latitude or longitude. Terrestrial surface magnetic anomalies (*e.g.*, iron deposits) produce local variations of 0.03–30 microtesla and are thus detectable.

Additional complications to these measurements include artificial magnetic field sources, the 0.01–0.1 microtesla solar daily variation, occasional magnetic storms producing erratic geomagnetic fluctuations of 0.01–5 microtesla, sudden-commencement ionospheric electrojets up to 0.3 microtesla, and various long-term secular variations in the geomagnetic field. Radio, television, and direct broadcast satellite signals probably are not detectable by individual nanodevices, but such detection may be indirectly achieved in vivo using nanorobot-accessible dedicated macroscale antennas implanted in the body.

Piezoelectric Stress Macrosensing

The piezoelectric effect is "the production of electrical polarization in a material by the application of mechanical stress". Many materials in the human body are piezoelectric, including tendon and elastin, dentin and bone. This polarization, or surface charge, varies as the physical stress imposed on the material changes over time.

Thus, measurement of the piezoelectric effect in bone or tendon can provide information on the physical loads that are being carried by these materials. For example, a shear stress applied along the long axis of a bone alters the polarization voltage that appears on a surface at right angles to the axis. In another experiment, piezoelectric surface charges on a loaded human femur were measured to range from –131 to +207 picocoulombs/cm^2 (–8 to +13 charges/micron^2) from one end of the bone to the other end, depending upon position along the shaft. Real-time monitoring of this data would permit specific inferences as to the amount of load the bone was carrying and from what direction, including bending, shearing, and twisting forces, from which whole-body activity states could subsequently be inferred. Knowledge of these surface charge variations might also be exploited in osteographic or functional navigation.

Optical Macrosensing

In vivo medical nanodevices can gather little useful optical information from the external environment. Here's why. In biological soft tissues, scattering dominates absorption except in the pigmented layers of the epidermis and stratum corneum. Thus the propagation of light in tissues may be regarded as occurring in two steps. In the first step, optical photons of intensity I_0 falling perpendicularly on skin are transmitted according to Beer's law through tissue to a depth z with a transmitted intensity of

$$I_z = I_0(1 - r_{sp})e^{\sigma_t z} \quad (\text{watts/m}^2)$$

where r_{sp} = specular reflection coefficient for visible light (Fresnel reflection at the air-tissue surface) ~ 4%–7%, or 0% if the original light source lies within the body; and the transmission coefficient $\sigma_t = \sigma_a$ (absorption coefficient ~ 300 m^{-1}) + σ_s (scattering coefficient ~30,000 m^{-1}) for various human soft tissues at optical wavelengths. (Coefficient values for the most heavily pigmented skin layers may be 5–7 times higher.)

Typical values for σ_t ~ 10,000–100,000 m^{-1}, average ~30,000 m^{-1} for soft tissue, although exceptionally clear tissues with σ_t = 1000 m^{-1} have been reported. Thus the mean free path of an optical photon in human soft tissue is 10–100 microns, average ~ 30 microns (~1.5 tissue cell-widths), up to an extreme maximum of ~1 mm for the most transparent tissues known.

For $\sigma_t \sim 30{,}000\ m^{-1}$, at z = 150 microns, $I_z/I_0 \sim 0.01$ and ~99% of all photons have been scattered at least once from their initial path.

In the second step, in an unbounded medium the patch of fully scattered photons continues to propagate through the tissue via diffusion until all photons are absorbed. The complicated governing diffusion equation has not yet been been completely solved analytically, but the asymptotic diffusive fluence is given roughly by

$$I_d - I_1 e^{\sigma_d z} \quad \text{(watts/m}^2\text{)}$$

where σ_d is the diffusion exponent or effective attenuation coefficient (averaging ~900 m^{-1} for typical soft tissues but with an extremely wide range reported over the ultraviolet, visible, and near-infrared wavelengths, from 10–1,000,000 m^{-1}.) As a crude approximation, the initial intensity of the fully scattered photon patch $I_i \sim I_0 (1-r_{sp})\, a_{scat}$, where $a_{scat} = \sigma_s/(\sigma_s + \sigma_a) \sim 0.987$ is the albedo for single particle scattering. Over the optical band from 400–700 nm, incident intensity I_0 = 100–400 watts/ m^2 when standing in direct sunlight; artificial lighting in homes and offices is typically 0.1–10 watts/m^2; moonlight provides only $I_0 \sim 10^{-4}$ watts/m^2; and the absolute threshold for human vision is ~10^{-8} watts/m^2.

For $\sigma_d \sim 900\ m^{-1}$, the intensity ratio of transmitted/incident visible light falls to $I_d/I_0 = 0.1$ at z = 2.5 mm depth (~eyelids), ~10^{-4} at z = 1 cm depth, and ~10^{-11} at z = 2.8 cm—a depth at which the tissue would be completely dark to the human eye even under direct sunlight illumination at the outermost skin surface. An optical nanosensor with N_{sensor} receiver elements, with each receiver element having an area A_e = 1 nm^2 and capable of single-photon detection, requires an integration time for reliable detection of e^{SNR} photons given by

$$t_{meas} - \frac{h v e^{SNR}}{I_d N_{sensor} A_e}$$

where h = 6.63 × 10^{-34} joule-sec and n = 4.3–7.5 × 10^{14} Hz for optical photons. Let us assume that a patient is standing in I_0 ~ 400 watt/m^2 direct unfiltered sunlight, and that t_{meas} = 1 sec,

SNR = 2, N_{sensor} = 25,000 elements giving a nanorobot eyespot of $A_e N_{sensor}$ = 0.025 micron^2, and that σ_d = 900 m^{-1}. Then the maximum tissue depth for optical photon detection is z_{max} ~ 17 mm (I_d ~ 10^{-4} watt/m^2) which includes a volume of tissue comprising up to ~30% of total body volume.

Changes in illumination at the minimum indoor artificial level of ~0.1 watts/m^2 are visible to z_{max} ~ 7 mm depth. This eyespot, if exposed to air on the outermost surface of the skin, is just sensitive enough to detect full moonlight.

Our general conclusion is that variations in normal indoor lighting may be directly measurable by nanorobots stationed within the outermost ~1 cm of body tissues. Imaging, as opposed to the simple illumination detection system described here, is a much more difficult design challenge.

NEURAL MACROSENSING

The ability to detect individual neural cell electrical discharges noninvasively in many different ways, coupled with the abilities (A) to recognize and identify specific desired target nerve cells and (B) to pool data gathered independently by spatially separated nanodevices in real time, offers the possibility of indirect neural macrosensing of complex environmental stimuli by eavesdropping on the body's own regular sensory signal traffic.

Nanomonitors positioned at the afferent nerve endings emanating from hair cells located in the otolithic membrane of the utricle and saccule and in the cristae ampullaris of the semicircular canals allow medical nanodevices to directly record, amplify, attenuate, or modulate the body's own sensations of gravity, rotation, and acceleration, although kinesthetic sensory management might also be required for complete control.

Motor neurons likewise can be monitored to keep track of limb motions and positions, or specific muscle activities, and even to exert control. Neuron-resident nanorobots may detect auditory effects induced in the brain by pulsed microwaves from external sources. Feline cochlear neurons sensitive to audible frequencies of >300 Hz respond to single microwave pulses at a threshold-specific absorption rate of 6–11 watts/kg-pulse, while in humans the threshold for effect, which depends on energy per pulse, may be as low as ~0.02 joule/m^2-pulse for people with low hearing threshold.

Olfactory and gustatory sensory neural traffic similarly may be eavesdropped by nanosensory instruments. Nerve taps in the medulla oblongata or at the phrenic nerve that drives the diaphragmatic muscles allow direct monitoring of respiratory activity. Pain signals may be recorded or modified as required, as can mechanical and temperature nerve impulses from other receptors located in the skin. Even psychological variables such as emotionality, vigilance, and mental workload may be directly monitored by measuring ANS activity in sympathetic efferent fibres outside of the brain.

The most complex and difficult challenge in neural macrosensing will be optic nerve taps. The retina is thoroughly vascularized, permitting ready access to both photoreceptor and integrator neurons. However, the optic nerve bundle itself has ~10^6 tightly bunched individual nerve fibres, a 10–100 MHz signal bandwidth, and significant natural data compression techniques which all must be untangled in real time. Developing algorithms capable of interpreting raw optical nerve traffic, say, to recognize a specific human face or a specific scene in the vision field, would prove a significant research challenge.

Direct monitoring of photo-receptors or retinal membrane potentials and other techniques may simplify untangling of the signal compression. Rapid visual field identification using artificial neural nets is also a subject of much current research interest. Eyeball rotations, eyelid position, pupil aperture, and lens accommodation under ciliary muscle control must be monitored to supplement the vision field analysis.

3

Molecular Nanotechnology

Modern surgery works only because cells have a remarkable ability to regroup, bury their dead and heal over the injury. Nanotechnology, "the manufacturing technology of the 21st century," should let us economically build a broad range of complex molecular machines (including, not incidentally, molecular computers). It will let us build fleets of computer controlled molecular tools much smaller than a human cell and built with the accuracy and precision of drug molecules. Such tools will let medicine, for the first time, intervene in a sophisticated and controlled way at the cellular and molecular level.

They could remove obstructions in the circulatory system, kill cancer cells, or take over the function of subcellular organelles. Just as today we have the artifical heart, so in the future we could have the artificial mitochondrion. Equally dramatic, nanotechnology will give us new instruments to examine tissue in unprecedented detail. Sensors smaller than a cell would give us an inside and exquisitely precise look at ongoing function.

Tissue that was either chemically fixed or flash frozen could be analysed literally down to the molecular level, giving a completely detailed "snapshot" of cellular, subcellular and molecular activities. There is broad agreement (though not consensus) that we will at some point in the future be able to inexpensively fabricate essentially any structure that is consistent with chemical and physical law and specified in molecular. The most direct route to achieving this capability involves positioning and assembling individual atoms and molecules in a fashion conceptually similar to snapping together LEGO blocks. By designing and building programmable self replicating manufacturing systems that incorporate these principles we should be able to achieve very low manufacturing costs. While the design and development of such programmable self replicating molecular manufacturing systems will be a major task and will likely require many years or a few decades, it appears that this kind of capability, to quote Feynman. Design concepts for general purpose self replicating manufacturing systems have been discussed for many years, and their utility in manufacturing has been emphasized recently. These proposals draw on a body of work started by von Neumann.

A wide range of methods have been considered. The von Neumann architecture for a self replicating system is the ancestral and archetypal proposal.

VON NEUMANN ARCHITECTURE

Von Neumann's proposal consisted of two central elements: a universal computer and a universal constructor. The universal computer contains a programme that directs the behaviour of the universal constructor. The universal constructor, in turn, is used to manufacture both another universal computer and another universal constructor. Once construction is finished the programme contained in the original universal computer is copied to the new universal computer and programme execution is started.

Von Neumann worked out the details for a constructor that worked in a theoretical two-dimensional cellular automata world (parts of his proposal have since been modeled computationally). The constructor had an arm which it could move about and which could be used to change the state of the cell at the tip of the arm.

By progressively sweeping the arm back and forth and changing the state of the cell at the tip, it was possible to create "objects" consisting of regions of the two-dimensional cellular automata world which were fully specified by the programme that controlled the constructor. While this solution demonstrates the theoretical validity of the idea, von Neumann's kinematic constructor (which was not worked out in such detail) has had perhaps a greater influence, for it is a model of general manufacturing which can more easily be adapted to the three-dimensional world in which we live. The kinematic constructor was a robotic arm which moved in three-space and which grasped parts from a sea of parts around it. These parts were then assembled into another kinematic constructor and its associated control computer.

An important point to notice is that self replication, while important, is not by itself an objective. A device able to make copies of itself but unable to make anything else would not be very valuable. Von Neumann's proposals centered around the combination of a universal constructor, which could make anything it was directed to make, and a universal computer, which could compute anything it was directed to compute. It is this ability to make any one of a broad range of structures under flexible programmatic control that is of value. The ability of the device to make copies of itself is simply a means to achieve low cost, rather than an end in itself.

DREXLER'S ARCHITECTURE

Drexler's assembler follows the von Neumann kinematic architecture, but is specialized for dealing with systems made of atoms. The emphasis here (in contrast to von Neumann's proposal) is on small size. The computer and constructor both shrink to the molecular scale, while the constructor takes on additional detail consistent with the desire to manipulate molecular structures with atomic precision.

The tip chemistry is logically similar to the ability of the von Neumann universal constructor to alter the state of a cell at the tip of the arm, but now the change in "state" corresponds to a change in molecular structure.

That is, we must specify a set of well defined chemical reactions that take place at the tip of the arm, and this set must be sufficient to allow the synthesis of the structures of interest. It is worth noting that current methods in computational chemistry are sufficient to model the kinds of structures that will appear in a broad class of molecular machines, including all of the structures and reactions needed for some assemblers

SIZES

Even including system overhead (power, connections, etc). the volume per element should still be less than 100 cubic nanometers. A 10,000 element logic system (enough to hold a small processor) would occupy a cube no more than 100 nanometers on a side.

That is, a volume only slightly larger than 0.001 cubic microns would be sufficient to hold a small computer. This compares favourably with the volume of a typical cell (thousands of cubic microns) and is even substantially smaller than subcellular organelles.

Operating continuously at a gigahertz such a computer would use less than 10^-9 watts. By comparison, the human body uses about 100 watts at rest and more during exercise. Slower operation and the use of reversible logic would reduce power consumption, quite possibly dramatically. A variety of molecular sensors and actuators would also fit in such a volume. A molecular "robotic arm" less than 100 nanometers long should be quite feasible, as well as molecular binding sites 10 nanometers in size or less.

By contrast, a single red blood cell is about 8 microns in diameter (over 80 times larger in linear dimensions than our 100 nanometer processor). Devices of the size range suggested above (~0.1 microns) would easily fit in the circulatory system and would even be able to enter individual cells.

KILLING OF CANCER CELLS

Given such molecular tools, we could design a small device able to identify and kill cancer cells. The device would have a small computer, several binding sites to determine the concentration of specific molecules, and a supply of some poison which could be selectively released and was able to kill a cell identified as cancerous. The device would circulate freely throughout the body, and would periodically sample its environment by determining whether the binding sites were or were not occupied.

Occupancy statistics would allow determination of concentration. Today's monoclonal antibodies are able to bind to only a single type of protein or other antigen, and have not proven effective against most cancers.

The cancer killing device suggested here could incorporate a dozen different binding sites and so could monitor the concentrations of a dozen different types of molecules. The computer could determine if the profile of concentrations fit a pre-programmed "cancerous" profile and would, when a cancerous profile was encountered, release the poison.

Beyond being able to determine the concentrations of different compounds, the cancer killer could also determine local pressure. A pressure sensor little more than 10 nanometers on a side would be sufficient to detect pressure changes of less than 0.1 atmospheres (a little over a pound per square inch. Linear increases in sensor volume would produce exponential increases in immunity to thermal noise and linear improvements in pressure sensitivity if that were to prove useful. Doubling the linear dimensions of the sensor would produce an eight-fold increase in both volume and pressure sensitivity). As acoustic signals in the megahertz range are commonly employed in diagnostics (ultrasound imaging of pregnant women, for example), the ability to detect such signals would permit the cancer killer to safely receive broadcast instructions. By using several macroscopic acoustic signal sources, the cancer killer could determine its location within the body much as a radio receiver on earth can use the transmissions from several satellites to determine its position (as in the widely used GPS system).

Megahertz transmission frequencies would also permit multiple samples of the pressure to be taken from the pressure sensor, as the CPU would be operating at gigahertz frequencies. The cancer killer could thus determine that it was located in (say) the big toe. If the objective was to kill a colon cancer, the cancer killer in the big toe would not release its poison. Very precise control over location of the cancer killer's activities could thus be achieved. The cancer killer could readily be reprogrammed to attack different targets (and could, in fact, be reprogrammed via acoustic signals transmitted while it was in the body). This general architecture could provide a flexible method of destroying unwanted structures (bacterial infestations, etc).

PROVIDING OXYGEN

A second application would be to provide metabolic support in the event of impaired circulation. Poor blood flow, caused by a variety of conditions, can result in serious tissue damage. A major cause of tissue damage is inadequate oxygen. A simple method of improving the levels of available oxygen despite reduced blood flow would be to provide an "artificial red blood cell." We will consider a simple design here: a sphere with an internal diameter of 0.1 microns (100 nanometers) filled with high pressure oxygen at ~1,000 atmospheres (about 10^8 pascals). The oxygen would be allowed to trickle out from the sphere at a constant rate (without feedback). Diamond has a Youngs modulus of about 10^*12* pascals. An atomically precise diamondoid structure should be able to tolerate a stress of greater than 5×10^{10} pascals (5% of the modulus).

Thus, a 0.1 micron sphere of oxygen at a pressure of 10^*8* pascals could be contained by a hollow diamondoid sphere with an internal diameter of 0.1 microns and a thickness of less than one nanometer. This thickness, thin as it is, results in an applied stress on the diamond of well under 1% of its modulus—from a purely structural point of view we should be able to use a very large "bucky ball," *i.e.*, a sphere whose surface is a single layer of graphite.

Perhaps the most complex issue involved in the selection of the material is the reaction of the body's immune system. While some suitable surface structure should exist which does not trigger a response by the immune system—after all, there are many surfaces in the body that are not attacked—the selection of a specific surface structure will require further research. To give a feeling for the range of possible surface structures, the hydrogenated diamond (111) surface could have a variety of "camouflauge" molecules covalently bound to its surface. A broad range of biological molecules could be anchored to the surface, either directly or via polymer tethers. The Van der Waals' equation of state is (p+a/v^2) (v–b) = RT, where p is the pressure, v is the volume per mole, R is the universal gas constant, T is the temperature in Kelvins, and a and b are constants specific to the particular gas involved. For oxygen, a = 1.36 atm liter^2/mole^2 and b = 0.03186 liter/mole and R = 0.0820568 liter-atmospheres/mole-kelvin. A mole of oxygen at 1,000 atmospheres and at body temperature (310 Kelvins) occupies 0.048 liters, or about 21 moles/liter. A mole of oxygen at 1 atmosphere and 310 Kelvins occupies 25.4 liters, or about 0.04 moles/liter.

This implies a compression of ~530 to 1. A resting human uses ~240 cc/minute of oxygen, so a liter of oxygen compressed to 1,000 atmospheres should be sufficient to maintain metabolism for about 36 hours (a day and a half). It might be desirable to replace less than a liter of blood with our microspheres of compressed oxygen, but it should still be quite feasible to provide oxygen to tissue even when circulation is severely compromised for periods of at least many hours from a single infusion.

Controlled release of oxygen from the diamondoid sphere could be done using the selective transport method proposed by Drexler and illustrated in figure. Transport in the "wrong" direction (for this application), but simply reversing the direction of rotor motion would result in transport from inside the reservoir to the external fluid. More sophisticated systems would release oxygen only when the measured external partial pressure of oxygen fell below a threshold level, and so could be used as an emergency reserve that would come into play only when normal circulation was (for some reason) interupted.

Full replacement of red blood cells would involve the design of devices able to absorb and compress oxygen when the partial pressure was above a high threshold (as in the lungs) while releasing it when the partial pressure was below a lower threshold (as in tissues using oxygen). In this case, selective transport of oxygen into an internal reservoir would be required. If a single stage did not provide a sufficiently selective transport system, a multi-staged or cascaded system could be used.

Compression of oxygen would presumably require a power system, perhaps taking energy from the combustion of glucose and oxygen (thus permitting free operation in tissue). Release of the compressed oxygen should allow recovery of a significant fraction of the energy used to compress it, so the total power consumed by such a device need not be great. If the device were to simultaneously absorb carbon dioxide when it was present at high concentrations (in the tissue) and release it when it was at low concentrations (in the lungs), then it would also provide a method of removing one of the major products of metabolic activity.

Calculations similar to those given above imply a human's oxygen intake and carbon dioxide output could both be handled for a period of about a day by about a liter of small spheres. As oxygen is being absorbed by our artificial red blood cells in the lungs at the same time that carbon dioxide is being released, and oxygen is being released in the tissues when carbon dioxide is being absorbed, the energy needed to compress one gas can be provided by decompressing the other.

The power system need only make up for losses caused by inefficiencies in this process. These losses could presumably be made small, thus allowing our artificial red blood cells to operate with little energy consumption. By comparison, a liter of blood normally contains ~0.2 liters of oxygen, while one liter of our spheres contained ~530 liters of oxygen (where "liter of oxygen" means, as is common in the literature on human oxygen consumption, one liter of the gas under standard conditions of temperature and pressure).

Thus, our spheres are over 2,000 times more efficient per unit volume than blood; taking into account that blood is only about half occupied by red blood cells, our spheres are over 1,000 times more efficient than red blood cells. Failure of a 0.1 micron sphere would result in creation of a bubble of oxygen less than 1 micron in diameter. Occasional failures could be tolerated. Given the extremely low defect rates projected for nanotechnology, such failures should be very infrequent.

ARTIFICIAL MITOCHONDRIA

In particular, the mitochondria will, at some point, fail. Increased oxygen levels in the presence of nonfunctional or partially functional mitochondria will be ineffective in restoring the tissue. However, more direct metabolic support could be provided. The direct release of ATP, coupled with selective release or absorption of critical metabolites (using the kind of selective transport system mentioned earlier), should be effective in restoring cellular function even when mitochondrial function had been compromised.

The devices restoring metabolite levels, injected into the body, should be able to operate autonomously for many hours (depending on power requirements, the storage capacity of the device and the release and uptake rates required to maintain metabolite levels). While levels of critical metabolites could be restored, other damage caused during the ischemic event would also have to be dealt with. In particular, there might have been significant free radical damage to various molecular structures within the cell, including its DNA.

If damage was significant restoring metabolite levels would be insufficient, by itself, to restore the cell to a healthy state. Various options could be pursued at this point. If the cellular condition was deteriorating (unchecked by the normal homeostatic mechanisms, which presumably would cease to function when cellular energy levels fell below a critical value), some general method of slowing further deterioration would be desirable.

Cooling of the tissue, or the injection of compounds that would slow or block deteriorative reactions would be desirable. As autonomous molecular machines with externally provided power could be used to restore function, maintaining function in the tissue itself would no longer be critical. Deliberately turning off the metabolism of the cell to prevent further damage would become a feasible option.

Following some interval of reduced (or even absent) metabolic activity during which damage was repaired, tissue metabolism could be restarted again in a controlled fashion. It is clear that this approach should be able to reverse substantially greater damage than can be dealt with today. A primary reason for this is that autonomous molecular machines using externally provided power would be able to continue operating even when the tissue itself was no longer functional. We would finally have an ability to heal injured cells, instead of simply helping injured cells to heal themselves.

RESEARCH

Advances in medical technology necessarily depend on our understanding of living systems. With the kind of devices discussed earlier, we should be able to explore and analyse living systems in greater detail than ever before considered possible. Autonomous molecular machines, operating in the human body, could monitor levels of different compounds and store that information in internal memory.

They could determine both their location and the time. Thus, information could be gathered about changing conditions inside the body, and that information could be tied to both the location and the time of collection. Physical samples of small volumes (nano tissue samples) could likewise be taken. These molecular machines could then be filtered out of the blood supply and the stored information (and samples) could be analysed. This would provide a picture of activities within healthy or injured tissue. This new knowledge would give us new insights and new approaches to curing the sick and healing the injured.

More dramatically, it should be feasible to take "snapshots" of tissue samples and analyse the structure down to the molecular level. First, a small tissue sample could be either fixed or frozen. Chemical fixation can be used to rapidly block most tissue changes. Ultra fast freezing of small tissue samples is an effective method of halting essentially all chemical processes and diffusion of all molecules. Once fixed or frozen, thc tissue sample could be analysed in a leisurely fashion. With nanotechnology (and indeed, to some extent with current STM and AFM technologies, though rather more expensively) it should be feasible to scan the tissue surface in molecular detail, and store that information in a computer.

Once the surface had been scanned, it could be removed in a very selective and precise fashion, and scanned again.

As an example, the use of a positionally controlled carbene has been proposed for use in the synthesis of complex diamondoid structures. Such a positionally controlled carbene is highly reactive and, if positioned at an appropriate site on the surface of the tissue being analysed, would readily react with a surface molecule. This surface molecule could then be removed. A wide variety of other "sticky" molecular tools could be brought up to the surface and allowed to react with surface molecules, which could then be removed, exposing the layers beneath. The use of a positionally controlled carbene implies that the environment in which it is used must be inert.

This requirement could be satisfied by analysing the tissue sample at very low temperature (a few Kelvins) and in a very good vacuum. Under these conditions the tissue specimen would remain stable during even a protracted analysis process. While this process can readily be envisioned for very small structures, nanotechnology should make massive parallelism feasible. That is, a single positional device could be used at a certain speed to provide information about a certain (rather small) volume of tissue in a reasonable time.

Nanotechnology should permit the manufacture of a large number of small devices, each able to analyse a small volume. Given enough such devices operating in parallel, larger volumes could be analysed and the information from many individual devices integrated to provide a coherent picture of the larger whole. Effective use of this option will require massive computational power—which will also be made feasible with nanotechnology. Estimates of the computational power that should be provided by nanotechnology exceed 10^24 logic operations per second for a single desktop computer. This amount of raw computational power should make control of a large number of parallel devices feasible, and should permit integration and analysis of the information so obtained. In short, tissue samples could be "frozen" (either literally by ultrafast cooling or figuratively by chemical fixation) and the entire resulting tissue sample could be analysed down to the level of individual molecules. The information so obtained could be processed by computers able to handle the flood of data produced. The resulting "snapshots" will provide us with an instantaneous look at metabolic and cellular activities across even relatively large volumes of tissue. Such an ability should revolutionize our understanding of the complex processes that take place in living systems.

The possibility of truly revolutionary advances in our medical abilities has also created renewed interest in cryonics. The abilities discussed here might well take years or decades to develop. It is quite natural to ask: "When might we see these systems actually used?" The scientifically correct answer is, of course, "We don't know." That said, it is worth noting that if progress in computer hardware continues as the trend lines of the last 50 years suggest, we should have some form of molecular manufacturing in the 2010 to 2020 time frame. After this, the medical applications will require some additional time to develop. The remarkably steady trend lines in computer hardware, however, give a false sense that there is a "schedule" and that developments will spontaneously happen at their appointed time.

CONTROL TO CHEMICAL SYNTHESIS

Manufactured products are made from atoms. The properties of those products depend on how those atoms are arranged. Viewed from the molecular level today's macroscopic manufacturing methods are crude and imprecise. Casting, milling, welding and all the other traditional manufacturing methods spray atoms about in great statistical herds. Even lithography (which already lets us put millions of transistors on a chip no bigger than your fingernail) is fundamentally statistical and random.

Exactly how many dopant atoms are in a single transistor and exactly where each individual dopant atom is located is neither specified nor known: if we have roughly the right number in roughly the right place, we can make a working transistor. For today, that is good enough.

The exception is chemistry. Large high purity crystals can have almost every atom in the right place. So, too, can many long polymers. The structures of proteins with hundreds and even thousands of amino acids can be specified down to the last atom. Most dramatically (and fortunately for us!) DNA strands with many tens of millions of bases can be copied with almost perfect accuracy. And it seems that almost any small molecule (with perhaps several dozens of atoms) can be synthesized, if only we have the skill and patience. Yet the laws of physics and chemistry in principle permit arranging and rearranging the elements in so many combinations and permutations that all of our manufacturing skills and all of our chemical skills barely suffice to scratch the surface of what is possible.

THE UTILITY OF DIAMOND

Almost any manufactured product could be improved, often by several orders of magnitude, if we could precisely control its structure at the molecular level. We often want our products to be light and strong. Diamond is light and strong: the strength-to-weight ratio of diamond is over 50 times that of steel. Yet we do not today have diamond spars in airplanes nor diamond hulls for rockets. Today we can't economically make diamond. Even if we could, simple diamond crystals can shatter. We'd have to modify the structure to make it tough and shatter proof: perhaps diamond fibres. While easily done in principle, we can't do this in practice today. Great strength and light weight are not the exclusive province of diamond: graphite can be stronger. And if we consider the many ways in which carbon atoms can be arranged and rearranged, then it's obvious that there are a host of other possibilities.

Yet all share a common problem: we can't yet economically make them in the exact shapes that we want. Great strength is only one property that we prize highly: when we make computers we are more concerned by electrical properties. Here, too, diamond excels.

Today's computers are made of semiconductors, and the semiconductor of choice is silicon. This is not because silicon is the ideal semiconductor from which to make computers, but because we know how to make devices from it. The computer industry has strong opinions about what makes a good logic device and what makes a good computer and diamond will let us make better computers than silicon. Diamond has a wider bandgap, hence electrical devices will work at higher temperatures. It has greater thermal conductivity, so devices can be more easily cooled. It has a greater breakdown field, hence devices can be smaller. It has higher electron and hole mobility which, when combined with higher electric fields, will result in higher speed.

But again, we see no diamond computers, just as we see no diamond airplanes: we can't economically manufacture them yet. Large pure crystals of silicon can be made relatively easily, but large pure crystals of diamond are scarce. We can etch the silicon surface and add dopants with a precision measured in tenths of microns, while the corresponding steps for diamond are more difficult. Not more difficult in principle: just more difficult today. Making computers highlights another problem. It's not enough to make a pure crystal, it must also have an extremely precise and complex pattern of impurities. The exact location of the dopant atoms in the semiconductor lattice controls how devices function and where signals can propagate.

Local order is crucial to make each device work, but long range complex order is crucial to make the computer as a whole work. While we can make some things today that are highly precise and have simple long range order (*e.g.*, crystals), it is the requirement for complex long range order that prevents us from making computers of the kind we'd like to make. While it's plausible we could make high density memory from crystals and perhaps some types of cellular automatons, we couldn't make anything that resembled the computers on the market today.

Today's high speed semiconductor-based digital computers (like the 80486 or the Pentium) have millions of logic elements wired together in complex and highly idiosyncratic patterns. This is well beyond the capabilities of crystal growth or bio-polymer synthesis. It will require a fundamentally new manufacturing technology: molecular manufacturing.

Today, there is a gap in our synthetic abilities: we can make complex mechanical machinery and electronic devices (including computers, which have millions of transistors), but we can't make such devices with the precision with which the chemist can synthesize a crystal, a bio-polymer, or a relatively small molecule. With chemistry we can make precise molecular structures and compounds, but we haven't been able to scale up that success to molecular computers (and other macroscopic products as precise as molecules).

Molecular manufacturing will, by definition, let us economically manufacture almost any specified structure that is consistent with the laws of chemistry and physics. To simplify the problem somewhat we can narrow our focus to structures that resemble diamond in a broad sense: the diamondoid structures as defined by Drexler. This class includes (among other things) diamond crystals of arbitrary shape but with stably terminated surfaces (*e.g.*, hydrogenated (111) or the like) and with impurities at precise locations in the diamond lattice (*e.g.*, substitutional boron). Our objective is to manufacture particular diamondoid structures once the location and type of every atom has been specified by design.

COMPUTER INDUSTRY

The attraction of molecular manufacturing for the computer industry should be clear. It should let us make computers at a manufacturing cost of less than a dollar per pound, operating at frequencies of tens of gigahertz or more, with linear dimensions for a single device of roughly 10 nanometers, high reliability, and energy dissipation (using conventional methods) of roughly 10^–18 joules per logic operation. If we make thermodynamically reversible computers (which the author and others have recently shown can be made from conventional electronic devices, *e.g.*, CMOS) then the energy dissipation per logic operation can be reduced to well below kT at T = 300 Kelvins (well below 10^–21 joules).

The computer industry is spending billions of dollars to make better computers. It is widely acknowledged within the industry that lithography is approaching its limits. Articles like *The Future of the Transistor, Miniaturization of Electronics and its Limits* and *Outlook for VLSI: Will the Balloon Burst?* Quite clearly show that conventional lithography will run out of steam (in perhaps a decade, though there is less agreement about the exact time frame). There is already interest in molecular logic devices and that interest will increase sharply as improvements in conventional manufacturing methods become increasingly difficult. However, any new proposal for manufacturing molecular computers will be weighed against (at least) the criteria mentioned above.

If it cannot easily beat conventional methods after they have been pushed to their uttermost limits, then it will be rejected. The computer industry will soon be pouring vast sums into research aimed at molecular computing, but the great bulk of funding will go towards well thought out proposals that offer a realistic possibility of substantially exceeding the performance of the ultimately evolved silicon VLSI technology that we expect to develop over the next decade. If you can't beat tomorrow's mainstream computers, you might as well not try. For this and many other reasons the class of diamondoid structures is a reasonable one to consider. The problem of building a diamondoid electronic computer captures many of the fundamental issues in molecular manufacturing, and poses clearly the issue of building large structures that cannot be made by regular repetition of some substructure (*e.g.*, the unit cell of a crystal or the monomeric unit in a bio-polymer).

This brings us to a core issue in molecular manufacturing: how do we synthesize such things? Today, we can synthesize diamond at low pressure and low temperature by using CVD (Chemical Vapour Deposition) methods.

Diamond CVD growth involves highly reactive species (radicals, carbenes, etc.) in a gas over the growing diamond surface that bombard and react with that surface at random. Because reaction sites are random, growth of many defect structures occurs (dislocations, etc.) as well as the desired perfect diamond structure.

Two fundamental mechanisms in the growth process include:

- Abstraction of hydrogens from the diamond surface leaving behind reactive sites (dangling bonds, radicals) and
- Interaction of carbon species (both reactive (CH2, CH3, etc.) as well as relatively unreactive species (C2H2)) with the surface, thus depositing carbon.

If we are to synthesize diamondoid structures it is plausible that we begin our search for the basic reaction steps involved in this synthesis by looking at existing reactions that occur in the CVD growth of diamond. The use of a reactive gas in the synthesis process, however, would seem to defeat any hope of making precisely patterned diamondoid structures, for the gas will interact with the growing surface at random.

POSITIONAL CONTROL IS FUNDAMENTAL

Here, we introduce the fundamental concept of molecular manufacturing: positional control over the site of reactions. To take a specific example we consider site specific hydrogen abstraction from the diamond (111) surface.

The ability to remove specific hydrogen atoms from the surface of the diamondoid work piece under construction is likely to be a fundamental unit operation in any attempt to make atomically precise diamondoid structures.

Hydrogen abstraction during CVD diamond growth typically involves a radical reaction between atomic H from the gas with H bonded to carbon on the surface producing H2. It is unclear how to make this process site specific. However, there are other structures with a high affinity for hydrogen which offer greater possibilities for positional control. In particular, the propynyl radical C3H3 has a great affinity for hydrogen.

Further, this radical has the very useful property that it has two ends: one end is a highly reactive radical while the other end is a stable sp3 carbon. Thus, we could synthesize a larger molecule with the propynyl radical at its end. The larger molecule would be held at the tip of a positional device.

The positional device would provide control over the orientation and position of this hydrogen abstraction tool (*e.g.*, a six degrees of freedom manipulator) and thus control the site of abstraction by controlling the position of the tool. Ab initio quantum chemical analysis of the abstraction of hydrogen from isobutane using an ethynyl abstraction tool supports the idea that the barrier to this reaction is zero. The reaction will proceed rapidly and, because of the large exothermicity, irreversibly.

Calculated barriers for abstraction from several other molecules were also small, suggesting that this hydrogen abstraction tool could be used to abstract hydrogen from a wide range of different molecules. The site specific abstraction of hydrogen illustrates the core concept in molecular manufacturing: selecting the reaction site by controlling the position and orientation of the reactants. The (relatively stiff) diamondoid workpiece is held in place, while a tool (in our example, a hydrogen abstraction tool) is positioned using a rather conventional (if also rather small) robotic arm. The ideas of using tools, controlling the position of those tools with a general purpose manipulator, and building complex structures by putting together components using those positionally controlled tools are rather common and even mundane at the macroscopic level.

At the molecular level, they are new and almost shocking: yet it is simply mapping onto the molecular world the concepts and ideas that have proven so useful and powerful in macroscopic manufacturing. By adding positional control we should be able to develop a method of molecular manufacturing which combines the best features of both conventional macroscopic manufacturing and chemical synthesis.

MOLECULAR TOOLS

If we are to grow diamond, we must also have carbon deposition tools. Drexler has suggested the use of positionally controlled carbenes and alkynes and proposed reaction pathways and surface structures where these tools would apply. In both cases, the tools are positioned at a precise point on the growing diamondoid structure and are used to deposit one or more carbon atoms at a desired location.

These deposition reactions parallel proposals in the CVD literature except for the addition of positional control (*e.g.*, at least one portion of the moiety must be part of an extended "handle" which can be held by a positional device). The broad range of possible tools coupled with the great power of ab initio computational chemistry should let us define and verify a complete set of molecular tools capable of synthesizing essentially any diamondoid structure.

TOOL USE

For such tools to be usable in a system context we must satisfy certain constraints. First and foremost, we must have a device capable of positioning the tool to within something like an atomic diameter. On the diamond (111) surface, the distance between adjacent hydrogens is about 2.5 Angstroms. Thus, positional accuracy of 1 to 2 angstroms for the hydrogen abstraction tool is required to prevent abstraction of the wrong hydrogen. Second, because the tools can be highly reactive, we require an inert environment. A simple inert environment is vacuum.

Compressed helium or some other inert gas would also work. Third, because it is the relative tool-workpiece position that must be controlled, the workpiece under construction must be relatively rigid (*e.g.*, not subject to vibrational motions that would exceed about an angstrom). Fourth and last, we must have some way of generating the sometimes highly reactive tools (*e.g.*, we need to define a precursor to the hydrogen abstraction tool, as well as precursors to the other tools).

In some sense, the analysis that we will now pursue is similar in type to retrosynthetic analysis. We start with the final product that we wish to build (a macroscopic diamondoid computer, for example), and then consider the possible predecessor structures which would yield the final product in one step. Then we consider the predecessor structures to those structures, and so on. We extend retrosynthetic analysis beyond its traditional bounds, but the general concept remains the same: given the finished product we deduce the possible ways in which it could have been constructed. This approach has been called "backwards chaining" by Drexler.

Work with SPMs (Scanning Probe Microscopes) clearly show that it is possible to achieve positional accuracies of a small fraction of an angstrom. Small (~0.1 microns) diamondoid "arms" or positional devices with similar positional accuracy are in principle quite feasible. The field of robotics provides a broad range of designs for positional devices which are largely scale independent. Shrinking these designs to submicron size is conceptually straightforwards. A factor of crucial importance in the design of molecular-scale positional devices is the accuracy with which the tip can be positioned, particularly in the face of thermal noise.

While atomically precise bearings and joints will not suffer from chatter, backlash, wear, tooth-to-tooth errors and other sources of inaccuracy caused by imprecise manufacturing, they will still suffer from positional errors caused by thermal noise. To control this source of error, it is essential that the robotic arm be very stiff, and so the use of stiff materials is desirable.

The Young's modulus for diamond is about 10^12 Pascals (very stiff), and back-of-the-envelope calculations show that a hollow cylinder of such material that is perhaps 100 nanometers long and 30 nanometers in diameter should have a positional accuracy at the tip, in the face of thermal noise at room temperature, of a small fraction of an atomic diameter.

A more detailed design and analysis of a jointed tubular robotic arm taking into account the bending and rocking motions of joints in the arm further supports this conclusion. Alternatives to the simple robotic arm are available which might be more attractive.

Other Requirements for Tool Use

Creating an inert environment also presents no

fundamental problems: high quality vacuums are common in laboratories today. If our objective is to have a very small very high quality vacuum, then a relatively thin wall of diamondoid material could be used as a barrier to keep a volume which was a modest fraction of a cubic micron free of any contaminants. If the volume were initially constructed free of contaminants then such a barrier would keep the inside free of any contaminants with high probability.

Because we are building diamondoid structures, they will be very stiff. As a consequence, it is relatively easy to meet the requirement that the objects that are being manufactured must themselves be stiff. Finally, generation of "activated" tools from relatively stable precursors can be done by a variety of methods. Because we are assuming an environment in which we have positional control we can use particularly simple precursors.

TRANSPORT ACROSS A BARRIER

Having introduced a diamondoid barrier to keep unwanted contaminants out (much as the bacterial wall allows bacteria to maintain an appropriate internal environment in the face of a fluctuating external environment) we must now solve the problem of getting desired raw materials through the barrier. We might, for example, wish to transport the hydrogen abstraction tool precursor across the diamondoid barrier. After use we will also need to eject the spent tool. Several ways to solve this problem are feasible. A proposal by Drexler is to use a rotor embedded in the diamondoid wall which moves binding sites from the outside of the wall to the inside of the wall.

By modulating the affinity of the binding site so that it will have high affinity for the desired molecule outside the barrier and low affinity inside the barrier, efficient transport across the barrier can be achieved. The desired molecule will bind to the binding site when it is outside the barrier, the binding site will be rotated to the inside of the barrier and the binding affinity reduced (in the illustrated proposal by mechanically pushing a rod into the binding site, thus physically precluding occupancy), and the molecule will be released on the inside of the barrier. The result is to increase the concentration of the desired molecule. A few stages of such a filtration system can achieve extremely high purities. The final stage, rather than ejecting the molecule into a liquid, would deliver the molecule into the inert internal environment in a well defined orientation where it could be further processed.

CONTROL SIGNALS

Finally, we will need a source of control signals for our molecular arm. One general approach would be to use a molecular computer. We will not consider a particular design for a molecular computer here, it is sufficient to note that many proposals for molecular computation have been considered in the literature and it is generally expected that some type of very small computational device will be feasible in a few decades. This completes our (all too brief) outline of a small device able to manufacture a broad class of diamondoid materials. Basically, the design is driven by the desire to provide the environment needed to synthesize diamond and diamondoid materials using the kinds of reactions that occur naturally during CVD growth of diamond. The device is itself made from diamondoid materials, which means that one such device can manufacture a second such device.

This ability to self replicate is crucial in achieving low manufacturing costs. As the reader might appreciate, the design and construction of one such general purpose device might well prove to be a time consuming and expensive undertaking. This cost cannot be justified unless the resulting device has great value.

If the device can self replicate then the successful design and development of one such device can be used to build an entirely new manufacturing technology. The manufacturing costs for the second, third, fourth.... 10^10.... etc. devices will consist largely of the raw materials and energy costs. Thus, a very large R&D cost can (if necessary) be justified.

The major driving force in the previous design was the desire to approach fundamental limits (strength, stiffness, thermal conductivity, electrical characteristics, etc.) in the manufactured products. This in turn implied the use of diamondoid materials, which, when coupled with the known chemistry of CVD diamond synthesis resulted in a high vacuum system with highly reactive intermediates.

Relaxing the materials requirements gives us a *much* wider range of possible structures. In particular, we can consider what is sometimes called "brick" or "building block" based nanotechnology. In this approach we first design a set of molecular building blocks, and then assemble the building blocks by the use of positional control.

The ribosome can be viewed as the prototype for this approach. The building blocks are amino acids, and they are linked together by the ribosome to form proteins.

Our approach differs in two principle respects: first, we add positional and orientational control over the building blocks in three dimensions, while the ribosome can only build structures that are fundamentally one dimensional (relying on linear structures that spontaneously fold into a particular shape to achieve a degree of control in three dimensions).

Second, rather than using relatively floppy polymers, we prefer relatively rigid bricks that can be bonded to each other in a stiff three-dimensional framework. This greatly simplifies the system. Indeed, with brick-based nanotechnology one can relatively easily envision the synthesis of a set of bricks that can, with the addition of positional control, be assembled into a wide range of structures with the stiffness of (say) wood.

A conceptually simple and relatively near-term way to achieve some degree of positional control would be to use conventional SPMs. The tips used in current SPM's are usually quite crude, and even when it is possible to make a very fine tip the range of possible structures is very limited (tungsten or some other simple material is generally all that's available). The design of tips for the SPM that incorporate individual molecules specifically synthesized for the purpose is a likely next step, and one that seems essential if we are to make progress in using SPM's to guide chemical reactions in a selective way. Such a "molecular manipulator" should be within reach of today's experimental technology. While the Young's modulus of the things it could make would be substantially inferior to that of diamond, this can be compensated by making them bigger. Scaling laws are such that increasing the size of an object by a factor of 10 also increases its stiffness by a factor of 10, and so reduces the positional inaccuracy from thermal vibration by a factor of 10.

If we wish to build a molecular "arm" out of bricks, then to achieve the same positional accuracy as with a diamondoid arm we will have to make the arm bigger—perhaps several tenths of a micron or more—but the basic design concepts that were discussed previously for use with a diamondoid arm still hold. We can still position the individual bricks accurately both in position and orientation, we can still build larger structures by putting together many bricks, we can still control the synthesis process by positional means, we can still make a positional device from bricks, etc.

Why pursue such an approach when it can only make relatively inferior materials?

There are three primary reasons:

1. It's easier to do
2. It could still make many things that are very valuable by today's standards and
3. Such systems could be used to make better systems (*e.g.*, diamondoid systems).

As the reader will appreciate, there are many possible candidates for bricks. Many researchers are already considering the design of molecular building blocks, although in most cases they do not consider positional control. Krummenacker discusses several of the issues surrounding the design of molecular bricks intended for assembly via positional control.

Adding positional control makes the synthesis of a broad range of molecular structures feasible, but at the same time requires the design and synthesis of a mechanism able to provide such control.

The obvious first step is to provide positional control using relatively modest extensions to today's SPMs. The most significant addition is the inclusion of a molecular tip, for today's SPMs typically have a limited range of possible tip structures and are often imprecise at the molecular scale.

4

Nanoresearch

CURRENT FINDINGS

MANUFACTURING

Molecular manufacturing is the use of programmable chemistry to build exponential manufacturing systems and high-performance products. There are several ways this can be achieved, each with its own benefits and drawbacks. This technology is coming soon—almost certainly within 20 years, and perhaps in less than a decade. When it arrives, it will come quickly. Molecular manufacturing can be built into a self-contained, tabletop factory that makes cheap products efficiently at molecular scale. The time from the first assembler to a flood of powerful and complex products may be less than a year. The potential benefits of such a technology are immense. Unfortunately, the risks are also immense.

PRODUCTS VALUE

Even a primitive diamond-building nanofactory can create products vastly more powerful than today's versions. Electrical power can be converted to motion, and *vice-versa,* with one-tenth the power loss and about 108(100,000,000) times more compactly. Computers can be a billion times smaller and use a million times less power. Materials can be about 100 times stronger than steel. This means that most human-scale products would consist almost entirely of empty space, reducing material requirements and cost. Most of the rest of the product would be structural, easy to design. Even the simplest products could be software-controlled at no extra hardware cost. Manufacturing of prototypes would be quite rapid-a few minutes to a few hours. Because manufacturing and prototyping are the same process, a successful prototype design could immediately be distributed for widespread use. A designer working with a few basic predesigned blocks could design, build, and test a simple product in less than a day. Products with complex interfaces to humans or to their surroundings-information appliances, automobiles, aerospace hardware, medical devices-would be limited by the time required to develop their software and test their functionality. However, in some fields the high time and money cost of manufacture slows other parts of the development cycle; this effect would disappear. An explosion of new, useful products could rapidly follow the widespread availability of a nanofactory.

ADVANTAGE OF CED NANOTECH

Manufacturing with nanotechnology can solve many of the world's current problems. Water shortage is a serious and growing problem. Most water is used for industry and agriculture; both of these requirements would be greatly reduced by products made by molecular manufacturing. Infectious disease is a continuing scourge in many parts of the world. Simple products like pipes, filters, and mosquito nets can greatly reduce this problem. Information and communication are valuable, but lacking in many places. Computers and display devices would become stunningly cheap. Electrical power is still not available in many areas. The efficient, cheap building of light, strong structures, electrical equipment, and power storage devices would allow the use of solar thermal power as a primary and abundant energy source. Environmental degradation is a serious problem worldwide. High-tech products can allow people to live with much less environmental impact. Many areas of the world cannot rapidly bootstrap a 20th century manufacturing infrastructure. Molecular manufacturing can be self-contained and clean; a single packing crate or suitcase could contain all equipment required for a village-scale industrial revolution. Finally, MNT will provide cheap and advanced equipment for medical research and health care, making improved medicine widely available. Much social unrest can be traced directly to material poverty, ill health, and ignorance. MNT can contribute to great reductions in all of these problems, and in the associated human suffering.

ADVANCED NANOTECH

Molecular nanotechnology will be a significant breakthrough, comparable perhaps to the Industrial Revolution-but compressed into a few years. This has the potential to disrupt many aspects of society and politics. The power of the technology may cause two competing nations to enter a disruptive and unstable arms race. Weapons and surveillance devices could be made small, cheap, powerful, and very numerous. Cheap manufacturing and duplication of designs could lead to economic upheaval. Overuse of inexpensive products could cause widespread environmental damage. Attempts to control these and other risks may lead to abusive restrictions, or create demand for a black market that would be very risky and almost impossible to stop; small nanofactories will be very easy to smuggle, and fully dangerous. There are numerous severe risks-including several different kinds of risk-that cannot all be prevented with the same approach. Simple, one-track solutions cannot work.

SIMPLISTIC REGULATION

Molecular nanotechnology manufacturing creates several severe risks, and each risk tempts a simple and extreme solution. However, a patchwork of extreme solutions will be both destructive and ineffective. For example, Bill Joyand others have proposed halting nanotechnology research entirely. This would not actually work; instead, it would relocate the research to less responsible venues. The risks might be delayed by a few years, but would be far worse when they appeared because the technology would be even less controllable. To take another example, economic upheaval might be prevented by strict commercial licensing of all uses of the technology. This has two problems. First, digital protection schemes for commercial products have often proved quite easy to crack. Second, if the technology is so restricted that it cannot disrupt existing economic systems, continuing poverty will kill millions of people each year, fueling backlash, social unrest, espionage, and independent development. Each risk must be reduced by some means that does not exacerbate others.

APPROACHES

Once molecular manufacturing is developed, it will have to be administered. There are several approaches that might help. CRN is not advocating any of these approaches at this point; we don't know enough about how the technology will be developed or in what context. We also have to point out that we don't think any one approach will be enough. Any effective program will require a balance of several different kinds of administration. Some possibilities include built-in technical restrictions in personal nanofactories; intellectual property reform; and international cooperation or monitoring of various kinds. Despite the difficulties and complexities, we believe that a solution can be found to preserve most of the potential benefits while avoiding the most severe risks.

DEVELOPING MNT

Early development of molecular nanotechnology increases some risks, but reduces others; overall, we think it's safest to develop as soon as possible. This is a preliminary conclusion, and we may change our opinion, but there are solid reasons for taking this position. The development of MNT seems inevitable sooner or later.

If development is delayed, it will rapidly become easier and cheaper, thus harder to control. Also, it's probably the case that early development will allow more time to develop MNT-based protective technologies—which may be necessary to cope with some dangerous MNT-based technologies. Finally, if it's done right, molecular manufacturing could save millions of lives per year and greatly decrease the environmental damage we're already doing. The costs of delay (opportunity costs) are significant, and may even outweigh the risks of development.

Even at this early stage, we can make some recommendations about how the technology of molecular manufacturing should be developed. Without some controls, advanced nanotechnology will probably be extremely dangerous—but desirable to many people. In addition, manufacturing systems will probably be portable and easy to duplicate. This means that it will be quite hard to control the use of the technology if unrestricted versions ever become widely available.

On the other hand, overly restrictive policy will encourage uncontrolled release. It seems likely that an early, closely guarded, international development program is probably the approach that retains the most control in the long run. CRN will continue working to clarify this issue and make specific recommendations.

ESSENTIAL STUDIES

Because of the largely unexpected transformational power of molecular manufacturing, it is urgent to understand theissues raised. To date, there has not been anything approaching an adequate study of these issues. CRN's recommended series of thirty essential studies is organized into five sections, covering fundamental theory, possible technological capabilities, bootstrapping potential, product capabilities, and policy questions. Several preliminary conclusions are stated, and because our understanding points to a crisis, a parallel process of conducting the studies is urged.

ESTIMATING TIMELINE

Molecular manufacturing (MM) means the ability to build devices, machines, and eventually whole products with every atom in its specified place. Today the theories for using mechanical chemistry to directly fabricate nanoscale structures are well-developed and awaiting progress in enabling technologies. Assuming all this theory works—and no one has established a problem with it yet—exponential general-purpose molecular manufacturing appears to be inevitable. It might become a reality by 2010 to 2015, more plausibly will by 2015 to 2020, and almost certainly will by 2020 to 2025. When it arrives, it will come quickly. MM can be built into a self-contained, personal factory (PN) that makes cheap products efficiently at molecular scale. The time from the first fabricator to a flood of powerful and complex products may be less than a year. The potential benefits of such a technology are immense. Unfortunately, the risks are also immense.

MOLECULAR MANUFACTURING

Again, this is probably done most easily by robotic assembly. Theoretical studies have shown that it should be possible to build diamond lattice by mechanically guided chemistry, ormechanochemistry. By building the lattice in various directions, a wide variety of parts can be made—parts that would be familiar to a mechanical engineer, such as levers and housings. A robotic system used to build the molecular parts could also be used to assemble the parts into a machine.

In fact, there is no reason why a robotic system can't build a copy of itself. In sharp contrast to conventional manufacturing, only a few (chemical) processes are needed to make any required shape. And with each atom in the right place, each manufactured part will be precisely the right size—so robotic assembly plans will be easy to program. A small nano-robotic device that can use supplied chemicals to manufacture nanoscaleproducts under external control is called a fabricator.

More than forty years ago, Richard Feynman said, "The principles of physics, as far as I can see, do not speak against the possibility of maneuvering things atom by atom." Molecular nanotechnology includes only one additional, and relatively easy, step: combining the small shapes and machines produced by individual chemical workstations into large products. The easiest way to do this is to combine small pieces into larger pieces, and then join those to make still larger ones.

This process is called *convergent assembly*, and it can be used to make products large enough to be used directly by people. CRN has published a peer-reviewed paper, titled "Design of a Primitive Nanofactory", showing how large numbers of fabricators can be combined to create a *personal nanofactory* (PN) capable of making human-scale products. It appears that this might be accomplished in as little as a few months after the first fabricator is built. The resulting PN would be easy to program to make a wide variety of products, including duplicate PNs.

Although there are several possible ways to develop an MM capability, the best way appears to be the creation of fabricators and then nanofactories that can make diamond lattice (as explained above). Diamond is very strong, and can be used to build a wide variety of useful gadgets including motors and computers. This implies that the products of a nanofactory will also be strong, and that active functionality can be extremely compact. For example, an engine powerful enough to drive a car would fill less than a cubic centimeter, and a modern supercomputer would require less than a cubic millimeter. Diamond structure would be at least ten times as strong as steel for the same weight—probably closer to 100 times as strong. Because of the simple, and massively parallel, manufacturing used by a nanofactory, the complexity of a product would not affect either the manufacturing cost or the time to build it.

A new design—any new design—could be built in just a few hours. A nanofactory, like an fabricator, will be able to duplicate itself. Nanofactories will be as cheap as any other product, so any desired number of nanofactories can be built. Since nanofactories can be used for final manufacturing as well as rapid prototyping, product design will not have to concern itself with "manufacturability." As soon as a prototype is designed, it can be built. As soon as the prototype is approved, mass production can be started—and finished a few hours later.

The design of an MM version of a product will actually be easier than today's process. Instead of designing a shape and then worrying about how to whittle down a block of material or carve out a mold, the designer simply specifies the shape—and the nanofactory will create diamond structure to fill the specified volume. Instead of worrying about fastening parts together, the designer can simply tell the CAD software that they should be attached. The surfaces to be joined will be covered by the CAD software with a simple mechanical interlocking mechanism (described in CRN's Nanofactory paper), and the convergent assembly process only needs to press them together. Because power and computer functionality will be much smaller than today's devices, the designer will have much less difficulty in making the functional parts of the design fit into the space required. And because a vast range of products can be specified by a single CAD system and manufactured by a single nanofactory design, a well-trained MNT designer will be able to design a large number of products, just as a well-trained software engineer can write a wide variety of programs.

The strength and power of products, the compactness of their functional components, and the ease and speed of design and production, combine to make MM a very useful technology. Vast amounts of money can be saved in the product design process, in manufacturing, in distribution and warehousing. New product lines can be designed, manufactured, and marketed in a few weeks. The same efficiencies apply to military hardware as well. Each new weapons system could be developed and deployed much more quickly and cheaply. Prototypes and tests would be generated much faster and cost far less. Since a prototype design could be immediately manufactured in any desired quantity, deployment would also be much faster. New kinds of weapon systems could be contemplated. Both commercial and military/governmental organizations will have a strong incentive to fund the rapid development of MM, even at a cost of billions of dollars.

FABRICATION

As fabricator is a small machine that can create precise

shapes out of molecules, assemble those shapes into machines, and ultimately duplicate itself when supplied with the necessary broadcast instruction stream. The duplication is necessary because a single fabricator could not build more than a small number of tiny products. A fabricator is a worthwhile goal, because although it can't make large products, many fabricators can be combined to form a nanofactory. CRN has published a echnical paper describing the process and techniques required to bootstrap from a sub-micron fabricator to a personal nanofactory; it appears that this can be done in a few months if suitable design and analysis is done beforehand. So we can assume that a fabricator project will include a nanofactory project, and that a useful nanofactory will appear within months of the first fabricator.

A wide range of products can be designed simply by sticking small functional blocks together; the joining process is covered in detail in the paper mentioned above. Effectively, then, the question of when we will see a flood of MM-built products boils down to the question of how quickly the first fabricator can be designed and built. Once the first desktop nanofactory has been built, its first product likely will be another identical nanofactory. Then, following the simple math of exponential duplication, it's easy to see that within months millions or even billions of personal nanofactories conceivably could be in operation. A key understanding of MM is that it leads not just to improved products, but to a vastly improved and accelerated means of production.

There is a difference between molecular manufacturing technology and today's nanomaterials research and other nanoscale technologies. Most of the nanotech work now being funded involves building small structures and searching for novel properties, then figuring out ways to use these new properties in new products. This is very useful work, and in many cases will be very profitable. But it is quite different from MM, which is concerned with building a single device: a flexible, easy-to-use, preferably large-scale, molecular manufacturing system. (Of course, once created this system could immediately start making a wide range of products.)

Some results of current nanotechnology research will be enabling technologies for MM: technologies that make it easier to build a fabricator. Non-nanotech fields will also contribute enabling technologies.

DESIGNING A FABRICATOR

Designing a fabricator will not be easy. Mechanochemistry, the formation or breaking of chemical bonds under direct mechanical control, has been demonstrated, but it will take a lot more work to develop the mechanochemical techniques to build diamond and other strong materials. These techniques will require some basic research; however, preliminary work (by Eric Drexler, Robert Freitas, Ralph Merkle, and John Michelsen, for example) shows that there are several different kinds of mechanochemical reactions that should be able to build diamond. Unless all this work is wrong *and* no other techniques can be discovered, building atomically precise diamondoid shapes will be possible. The small-scale robotic device to do the required mechanochemical operations has to be designed, including the control system. This is mostly a matter of simple mechanics. The integration of the mechanochemical device with other devices to support the parts and product, deliver "feedstock" chemicals from an uncontrolled exterior to a well-controlled interior, and so on should also be relatively straightforward—at least compared with designing a spacecraft.

A modern spacecraft contains millions of parts (estimates for the Space Shuttle range from 2.5 to six million). A large spacecraft design must account for fluid dynamics, aerodynamics, vibration and resonance on many time scales, avionics and other control, chemical engineering, mechanical engineering, electrical engineering, combustion dynamics, hydraulics, cryogenics, and biomedical issues. (Thanks to an anonymous poster on Slashdot for pointing this out.) By contrast, a fabricator design must account for chemistry, mechanical engineering including stiffness, control structures, and a different set of forces than we're used to at the macroscale (*e.g.* van der Waals force). Note that many problems can be treated as mechanical engineering issues without greatly increasing the size and complexity of the fabricator. One example is thermal noise: as analysed in *Nanosystems,* if the parts are stiff enough, it's not a problem even at room temperature.

BUILDING FABRICATOR

uilding the first fabricator may be even harder than designing it. (Building the second and subsequent fabricators will be relatively easy.) If the first fabricator is diamond-based, the diamond must be formed in small precise shapes without the benefit of fabricator mechanisms. If the first fabricator is built of DNA, protein, or other "wet" chemistry products, it must either work underwater while protecting the workpiece, or must work after being dried. Neither of these option is very attractive.

However, we are already learning to do mechano-chemistry and nanomanipulation with scanning probe microscopes. The use of buckytubes as scanning probes is fairly new, but is already proving useful. There are a variety of potential ways to build structures even smaller and more precise to do the required chemistry. Again, unless every single possibility we can think of turns out to be unfeasible, a fabricator can be built.

FABRICATOR DESIGN

If a fabricator project is not feasible today, it will surely be feasible in a few years. Most of the enabling technologies mentioned here, and many others as well, are being actively developed for their present-day commercial potential. As the technologies develop, they will reach a point where they can easily be re-used in a fabricator project. The mechanics of the project will become far easier in just a few years. The chemistry will become easier as more powerful computers are developed for simulation, but already it is feasible to test individual reactions in simulation. The question is not whether a fabricator project is feasible, but when it will become economically viable or a military necessity.

A new, large spacecraft or weapon system costs tens of billions of US$ to develop, and molecular nanotechnology will be far more useful than any single aerospace or weapons system. In today's dollars, total development cost for the original Space Shuttle was probably around $10-15 billion. At that rate, each part would have cost an average of $2,000-$6,000 to design. How many parts will a fabricator require? Estimates of the atom count, based in part on comparisons with bacteria, frequently come in around 1 billion atoms. Diamond has 176 carbon atoms per cubic nanometer, so if each part were only one cubic nanometer, a fabricator might have 6 million parts—comparable to the Shuttle. With parts 10 nanometers on a side, it would have only 6,000 parts. For comparison, a typical four-cylinder automobile engine has about 450 parts and a bacterium may have 3,600 different molecules. As opposed to a "wet" design like a bacterium or a cutting-edge aerospace design, most of a fabricator's parts would not interact with each other and could be designed separately. It appears, then, that design of a fabricator falls somewhere between a car engine and the Space Shuttle in complexity. Construction, if not feasible today, will be feasible soon.

The Space Shuttle took less than ten years to design and build, from 1972 to 1981. The atomic bomb took only three years, from 1942 to 1945. Both of these programs involved more new science research and more development of new technologies and techniques than an assembler program would likely require. As analysed above, they probably cost more too. The main question in estimating a timeline for fabricator development, then, is when it will be technically and politically feasible. There are probably five or more nations, and perhaps several large companies, that could finance a molecular fabricator effort starting in this decade. The technical feasibility depends on the enabling technologies. Even a single present-day technology, dip-pen nanolithography, may be able to fabricate an entire proto-fabricator with sufficient effort. At this point, we have not seen anything to make us believe that a five-year $10 billion fabricator project, starting today, would be infeasible, though we don't yet know enough to estimate its chance of success. Five years from now, we expect that a five-year project will be obviously feasible, and its cost may be well under $5 billion.

The National Science Foundation, and others, have estimated that even non-MM nanotechnology will be worth a trillion dollars or more by 2015. By the time people realise that it's possible to build a nano-based manufacturing system, it will probably be obvious that such a project would be quite profitable (in addition to the military imperatives). This implies that companies and/or governments will start crash programs, comparable perhaps to the Manhattan project. Of course there are other development scenarios, but we feel this is one of the more likely ones. We also cannot rule out the possibility that a large, well-funded, secret development program for molecular manufacturing has been in operation somewhere for several years and may achieve success sooner than any public program.

The Threshold of Nanotechnology

In the last chapter, we looked at the state of current research, but from there to thenanotechnology of even the Pocket Library scenario is a leap. How will this gap be crossed? In this chapter, we outline how emerging technologies can lead to nanotechnology. The actual path to nanotechnology-the one that history books will record-could emerge from any one of the research directions in physics, biochemistry, and chemistry recounted in the last chapter, or (more likely) from a combination of them. The availability of so many good options builds confidence that the goal can be reached, even while it decreases confidence that some particular path will be fastest. To see how advances might cross the gap from present technology to early nanotechnology, let's follow one path out of the many possible.

Bridging the Gap

One way to bridge the gap would through the development of an AFM-based *molecular manipulator* capable of doing primitive molecular manufacturing. This device would combine a simple molecular device—a molecular gripper—with an AFM positioning mechanism. An AFM can move its tip with precision; a molecular manipulator would add a gripper to the tip to hold a molecular tool. A molecular manipulator of this kind would guide chemical reactions by positioning molecules, like a slow, simple, but enormous assembler. (In our standard simulation view, where a molecular assembler arm fits in a room, the AFM apparatus of a molecular manipulator would be the size of a moon.) Despite its limits, an AFM molecular manipulator will be a striking advance.

Scenario: Developing a Molecular Manipulator

Several years ago, researchers at the University of Brobdingnag began work on developing a molecular manipulator. To reach this goal, a team of a dozen physicists, chemists, and protein researchers banded together (some working full time, some part time) and began the creative teamwork needed to solve the basic problems.

First they needed to attach a gripper to an AFM tip. As grippers, they chose fragments of antibody molecules, the selectively sticky proteins that the immune system uses to bind and identify germs. If they could get the "back" of the molecule stuck onto a tip, then the "front" could bind and hold molecular tools. (The advantage of antibody fragments was this: freedom of tool choice.

Since the late 1980s, researchers had been able to generate antibodies able to bind almost any preselected molecule-or molecular tool.) In parallel, the U. Brob AFM researchers worked on placing tips in a precise location and then holding them there with atomic accuracy for seconds at a time. This proved straightforward. They used techniques developed elsewhere during the early 1990s, adding only modest refinements.

They now had their gripper and a way of putting it where they wanted it, but they needed a set of tools. The gripper was like the chuck of a drill, waiting to have different bits fitted into its tool-holder slot. So as the final step, the synthetic chemists on the team made a dozen different molecular tools, all identical at one end but different at the other. The similar parts all bound to the same antibody tool-holder, slotting neatly into position. The different parts were all chemically reactive in different ways. Like the molecular tools in the hall of assembler arms in Chapter 3, each of these tools could use a chemical reaction to transfer some atoms to a molecular object under construction.

Developing the molecular tool kit was the toughest part of the project; it took about as much work as had gone into duplicating the palytoxin molecule back in the 1980s. None of the tasks in the project demanded the solution of a deep scientific puzzle, and none demanded the solution of a notoriously difficult engineering problem. Each task had many possible solutions, the problem was to find a compatible set of solutions and apply them. After a few years, the solutions came together and the U. Brob research team began building new molecules by molecular manipulation. Now many teams are doing likewise.

MOLECULAR GRIPPERS TOOLS

To build something with the U. Brob team's AFM-based molecular manipulator system, you use it as follows: First, choose a surface to build on and place it under the tip in a pool of liquid. Then dunk the AFM tip into the liquid, bringing it down to the surface, and back it off a little. Construction can now begin as soon as a tool is loaded into the gripper.

Tubes and pumps can flow different liquids over the surface and past the gripper, carrying different tool molecules. If you want to do something with a tool of Type A, you wash in the proper liquid, and a Type A molecule promptly sticks. Once it is in the gripper, you can use the AFM mechanism to move it around and put it where you want it. Move it up to the surface at a convenient spot, wait a few seconds, and it reacts, forming a bond and leaving a molecular fragment attached to the spot you chose. To add a different fragment, you can use a tool of Type B: you back up the tip, flow in a fresh liquid carrying the new tools, and in a moment a tool of the new type is bound in place and ready to apply, either on or alongside the first spot. Step by step, you build up a precise molecular structure.

Each step takes only seconds. Molecular tools pop into the gripper in a fraction of a second, and used tools pop off at the same rate. Once the tip has positioned a molecule, it reacts quickly, about a million times faster than unwanted reactions at other sites. In this way, the molecular manipulator gives good control of where reactions will occur (though it is not as reliable as an advanced assembler would be). It is fairly fast by a chemist's standards—per cycle—but still a million times slower than an advanced assembler. It can perform a variety of steps, but isn't as flexible and capable as an advanced assembler. In short, it is hardly the last word in nanotechnology, yet is a great advance over what has gone before.

Products

With its ability to accelerate desired reactions by a factor of a million or so, the U. Brob team's molecular manipulator can perform 10,000 to 100,000 steps with good reliability. Back in the 1980s, chemists making protein molecules struggled to perform just one hundred steps.

The U. Brob research team (and its many imitators) can now build structures that are stronger and easier to design than proteins: not floppy, folded chains, but rugged objects held together by a sturdy network of bonds. Though not as strong and dense as diamond, these structures are like bits of a tough engineering plastic.

A specially adapted computer-aided design system makes it easy to design molecular objects made from these materials. Yet the AFM-based molecular manipulator has one grave disadvantage: It does chemistry one molecule at a time, and it ties up a machine as expensive as a car for hours or days to produce that one large molecule. Some molecules, though, are valuable enough to be worth building even one at a time. These draw prompt attention.

A single molecule isn't much use as a dye, a drug, or a

floor wax, but it can have substantial value if it provides useful information. The U. Brob team quickly publishes a pile of scientific papers based on experiments with single molecules: they build a molecule, probe it, report the results, and build another. Some of these results show chemists elsewhere in the multibillion-dollar chemical industry how to design new catalysts, molecules that can help make other molecules more cheaply, cleanly, and efficiently.

This information is worth a lot. Three new products of special interest are among the first to be made. The first-molecular electronics-begins with experiments conducted by a research group at a computer chip company. They use their molecular manipulator to build single molecules and probe them, gradually learning how to build the parts needed for molecular electronic computers.

These new computers don't immediately become practical, because the costs are too high for making such large molecules with AFM-based technology. Yet some companies begin to produce simpler molecular electronic devices for use in sensors and specialized high-speed signal processing. A specialty industry is born and begins to expand. The second product is a gene reader, a complex molecular device built on the surface of a chip. The biologists who built the reader combined proteins borrowed from cells with special-purpose molecular machines designed from scratch. The result was a molecular system that binds DNA molecules and pulls them past a read-head-like tape through a tape recorder. The device works as fast as some naturally occurring molecular machines that read DNA, with one key advantage: it outputs its data electronically. At that speed, a single device can read a human genome in about a year. Though still too expensive for a doctor's office, these readers are promptly in great demand from research laboratories. Another small industry is born.

The third product is far more important, in the long run: replacement tips for molecular manipulators, grippers, and tools that are better than the originals. With these new, more versatile devices, researchers are now building more ambitious products and tools.

More Scenario: The Next Step to Nanotechnology

While the physicist-led team at U. Brob was finishing its work on the AFM-based molecular manipulator, a chemist-led team at the University of Lilliput was working furiously. They saw the U. Brob desktop machine as too large and its expected products as too expensive. Even back in the 1980s, David Biegelsen of the Xerox Palo Alto Research Center had noted, "The main drawback I see to using a hybrid protoassembler is that it would take a long time to build just one unit. Building requires a series of atom-by-atom construction steps. It would be better to build in parallel from the very beginning, making many trillions of these molecules all at the same time. I think there is tremendous power in parallel assembly. Maybe another field, chemistry or biology, offers a better way to do it." The chemists at U. Lill aimed to develop that better way, building first simple and then more and more complex molecular machines. The eventual result was a primitive molecular assembler able to build molecular objects by the trillions.

How did the chemists achieve this? During the years when the U. Brob team was developing the molecular manipulator, researchers working in protein science and synthetic chemistry had made better and better systems of molecular building blocks. Chemists were well prepared for doing this: by the late 1980s, it had become possible to build stable structures the size of medium-sized protein molecules, and work had begun to focus on making these molecules perform useful work by binding and modifying other molecules. Chemists learned to use these sophisticated catalysts-early molecular devices-to make their own work easier by helping in the manufacture of still more large molecules.

Another traditional chemist's tool was software for doing computer-aided design. The early software designed by Jay Ponder and Frederic Richards of Yale University ultimately led to semi-automatic tools for designing molecules of a particular shape and function. Chemists then could easily design molecules that would self-assemble into larger structures, several tens of nanometers across.

Their primitive assemblers built much the same sorts of products that the U. Brob molecular manipulator did; the tools were similar, and speed and accuracy were about the same. Yet there was one dramatic advantage: About 1,000,000,000, 000,000,000,000 U. Lill assemblers could fit in the space occupied by one U. Brob manipulator, and it was easy to produce a mere 1,000,000,000,000,000 times as much product at the same cost. With the first, primitive assemblers, construction was slow because each step required new liquid baths and several seconds of soaking and waiting, and a typical product might take thousands of steps. Nonetheless, the U. Lill team made a lot of money licensing their technology to researchers trying to commercialize products they had first researched with the U. Brob machine.

After starting an independent company (Nanofabricators, Inc.), they poured their research efforts into building better machines. Within a few years, they had assemblers with multiple grippers, each loaded with a different kind of tool; flashes of coloured light would flip molecules from state to state (they copied these molecules from the pigments of the retina of the eye); flipping molecules would change tools and change rod positions. Soaking and waiting become a thing of the past, and soon they were pouring out parts that, when mixed with liquid and added to dishes with special blank chips would build up the dense memory layers that made possible the Pocket Library.

That was when things started moving fast. The semiconductor industry went the way of the vacuum tube industry. Money and talent poured into the new technology. Molecular CAD tools got better, assemblers made it easy to build what was designed, and fast production and testing made molecular engineering as easy as playing with software. Assemblers got better, faster, and cheaper. Researchers used assemblers to build nanocomputers, and nanocomputers to control better, faster assemblers. Using tools to build better tools is an ancient story.

Will developments in the late pre-breakthrough days be as just described? Certainly not: the technical approaches will differ, and the U.S. academic research setting implied by the scenario could easily be replaced by academic, commercial, governmental, or military research settings in any of the advanced nations. What do seem realistic are the implied requirements for effort, technology, and time, as well as the basic capabilities of different devices. We are approaching a threshold of capability beyond which further advances will become easy and fast.

NANOPARTICLES

The rapidly developing nanoscience and nanotechnology reveal many exciting features of materials at the nanometer (10^{-9} meter) scale, and the possibility of manipulation of such features to create novel materials or products that are previously unthinkable. Many believe that nanoscience and nanotechnology development will become a new industrial revolution.

On the other hand, together with the numerous beneficial potentials, scientists also warn against possible adverse health and environmental impacts brought by the vastly unknown physicochemical properties of materials at the nanoscale. Among the different environment, health and safety issues, nanomaterials, in particular, nanoparticles probably cause the highest concern because of their potential of rapid uptake by biological system, unknown distribution pathways and potential interactions with various components of the biological system.

OBSERVATION

Nanoparticles are previously known in the air pollution field as ultrafine particles, which are defined as particles less than 100 nm in aerodynamic equivalent diameter (AED). Some ultrafine particles are generated by natural processes such as volcanic activities and wild fires. Others are derived incidentally from anthropogenic sources, *e.g.* engines and incinerators. In general, high energy activities involving organic or inorganic all have the potentials of producing ultrafine particles as one of the byproducts. On the other hand, with the development of nanoscience and nanotechnology, many nanoparticles are purposely made for particular functions. These are usually called engineered nanoparticles, and are defined as engineered particles with at least one dimension smaller than 100 nm.

Nanoparticles have unique physicochemical properties that are not found in their parent materials. In general, they have much higher reactivity, and because of their ultra small size, they can easily penetrate skin or cells, rapidly distribute in human body, and even directly interact with organelles within cells. Their huge surface area to mass ratio increases the chemical activities and therefore allows them to become efficient catalysts. These increased chemical and biological activities result in many engineered nanoparticles being designed for specific purposes, including diagnostic or therapeutic medical uses and environmental remediation that require introduction of these novel materials into human body or the environment.

The novel properties of nanomaterials, however, are a two-sided sword. The same properties that allow the wonderful beneficial uses also imply potentially devastating adverse health effects. Rapid uptake through skin and epithelial cells, capability to translocate along neurons, plus the known toxicity of some parent materials warrant careful environment, health and safety evaluation side by side with every nanomaterial application study.

NANOMATERIAL TOXICITY

Researchers worldwide are only beginning to look into toxicity of various nanomaterials. Knowledge of this newly emerged field, sometimes called "Nanotoxicity", is continuously being generated. The following is a discussion of certain known toxicity of several common groups of nanomaterials.

Carbonaceous nanomaterials-Fullerenes, consist of repeating hexagonal and pentagonal rings of carbon atoms. They include buckyballs (C_{60}) and other spherical structures, and single wall and multi-wall carbon nanotubes. Buckyballs are extremely stable. They can trap other atom or molecule inside the carbon structure. Possible applications include superconducting material, drug delivery system, lubricant and catalyst. Carbon nanotubes can also be superconducting, and can be made into mechanically super-strong materials which are 100 times the strength, but one sixth the weight of steel, among many other potential applications.

Carbon has always been considered a non-toxic element, and not very water soluble. However, recent studies found that buckyballs have considerable water solubility, and can cause lipid peroxidation in fish brain after 48 hours at a concentration of 500 parts per billion (ppb). This kind of oxidative stress is usually a first sign of biological damage. Single-wall nanotube has been found to cause other oxidative stresses such as free radical formation and depletion of antioxidants in cells. Metal impurity in synthesis of nanotube may also add to the toxicity. Besides, physical dimensions of some nanotubes may cause pulmonary fibrosis and cancer similar to the effect of asbestos.

Another example of potential toxicity is quantum dots (Q-Dots), which are nanoparticles of semiconducting materials. Their sizes are so small that they interfere with quantum behaviour of atoms. One of the results is they display tunable, based on particle size, absorption and emission spectrum over visible wavelength. This property gives quantum dots a strong potential in medical imaging. However, the current quantum dots are made of a cadmium-selenium (CdSe) core with a zinc sulfide (ZnS) shell, and all three metals are known to be toxic to different extents. The high surface area and rapid distribution of the ultra small particles are expected to further escalate the potential toxic effects once introduced into a biological system.

Some nanomaterials have already found their ways into commercial products, such as titanium oxide (TiO_2) for photochemical oxidation, self-cleaning paints, sunscreen products, etc.; and zinc oxide (ZnO) in cosmetics and sunscreen products. Other nanoparticles are being used in specialized applications, *e.g.* iron particles for oxidation of contaminants in groundwater, copper oxide (CuO) for antimicrobial activities; and iron oxide (Fe_2O_3) as a contrast agent for magnetic resonance imaging. Although most of these parent materials are not considered to be significantly toxic, the ultra small size of the particles again may create unknown toxicity or environmental impacts. There were already studies that showed the toxicity of nanosize TiO_2increases proportionally to the surface area, not the mass as in conventional toxicology. A recent study also measured free nanosize TiO_2 particles off the surface applied with a paint containing the nanoparticles.

OCCUPATIONAL HAZARDS OF NANOMATERIALS

The environmental, health and safety concerns of nanomaterials in fact span from research safety, manufacturing safety, to consumer (product) safety, and include environmental impacts throughout the life cycle, until final disposal of the nanomaterial products and wastes. Among these concerns, research and pilot scale production safety is most relevant to the research and development work at HKUST. Due to the rapid development and many uncharted areas in nanoscience and nanotechnology, the practice of occupational health and safety also faces many challenges. As for many conventional workplace contaminants, airborne route is considered the most crucial for worker protection.

The shape and size of airborne nanoparticles, and the conglomeration status are some of the main factors determining the hazard. Nanoparticles in liquid phase may present skin hazard when contacted.

The hazard of the liquid matrix, its potential interaction with the nanoparticles, *e.g.* serving as a carrier that facilitates skin penetration, and the conglomeration status of the particles will affect the overall hazard. In either the airborne or liquid phase, the nanomatierals have the potential to adversely impact the environment if they are not properly contained during usage and disposed of afterwards.

CHALLENGES IN ASSESSING AND CONTROLLING

A major challenge in evaluating hazard posed by airborne nanoparticles is to find a (set of) suitable exposure metric(s).

It will be necessary to consider number concentration and surface area, other than the conventional mass concentration, then appropriate sampling and analytical methods, and eventually health-based standards, will have to be developed. Quite a few number concentration direct reading instruments originally designed for ultrafine particle monitoring are available, but their applicability to monitoring of nanoparticles in a work environment remains to be tested. Also the reliability of the detection technology and how "direct" the reading really is are also main concerns to be addressed. In order to differentiate engineered nanoparticles of interest with natural or incidental nanosized particles that may be present in the background, examination by transmission electron microscopy (TEM) remains the golden standard for characterizing and enumerating nanoparticles.

At least one passive sampling method coupled with enumeration by electron microscope has been developed, its precision and accuracy are yet to be determined. Governmental agencies of developed countries such as the National Institute for Occupational Safety and Health (NIOSH) of the US, and Health and Safety Executive (HSE) of the UK, are actively working to fill up these knowledge gaps in assessing hazards of airborne nanoparticles.

Another major challenge is control technologies. Traditionally, high efficiency particulate air (HEPA) filters are used for removing airborne particulates. However, before the rise of nanomaterial research, it has never been proved whether HEPA filter can capture nano scale particles. Theoretically the filtering efficiency of HEPA increases with decreasing size below 0.1-0.3 μm, which is the most difficult size range to capture according to physics principles of filtration by fibrous materials. Previously there were only measured data points down to 0.01 μm (10 nm), beyond which the filtration efficiency was by extrapolation.

However, according to preliminary data presented in the Second International Symposium on Nanotechnology and Occupational Health (Minneapolis, October 2005) by a group of researchers from University of Minnesota, 3M and NIOSH,

HEPA filter is capable of removing more than 99.97% of particles down to the range of 2-4 nm. No significant thermal rebound was observed. This means with our current knowledge, HEPA filter can be deployed as an effective means to contain and control airborne nanoparticles at least down to the low nm size range.

As for skin protection, most of the glove and protective clothing materials currently in use have not been tested against nanosize particles. Presumably the materials should be protective against nanoparticle conglomerates. Some suggested treating airborne nanoparticles as toxic gas in terms of personnel protection. In any case, until information on interaction of nanoparticles and protective material is available, skin protection should at least be adequate to prevent penetration of the carrier medium of nanoparticles in the liquid phase.

Either in air or liquid phase, nanoparticles tend to agglomerate instead of existing in single particle form. This is helpful from a hazard control point of view. However, certain future applications or manufacturing processes may rely on dispersed nanoparticles. Therefore it is prudent to plan safety measures assuming the worst case scenario.

Some basic precautionary measures have been suggested by organizations such as HSE of UK and NIOSH of US. These include treating all nanomaterials that may possess unique physical chemical properties as "new chemicals" for regulatory purpose. Material and safety information raw material can only be viewed, at best, as a starting point or reference for assessing the risk of the nanomaterial product. A good example is carbon, which is practically non-toxic as a parent material; but a group of scientists have already established a toxicity level of oral lethal dose 50 (LD50) of 300 mg/kg for C_{60} buckyballs, which is considered moderately toxic.

An important concept that worth mentioning is "Responsible Nanotechnology", which basically says, there are many promising applications for nanoscience and nanotechnology, however, there are also uncertainties in potential health and environmental effects. Therefore we should not repeat the errors we made before in cases like DDT and chlorofluorocarbons, where people focused on the beneficial use of the new chemicals so much that they did not carefully check for potential health and environmental impacts until damages were done. Concurrent development of technology and environment, health and safety assessment and controls must go hand in hand in this potentially revolutionary field of research.

The UK Royal Society and Royal Academy of Engineering published a report titled "Nanoscience and nanotechnologies: opportunities and uncertainties" in July 2004.

They proposed the following general guidelines, which form a good foundation for responsible nanotechnology development:

- Treat nanomaterials as if they are hazardous
- Seek to remove them from waste stream
- Don't apply free nanomaterials to environment unless benefit clearly outweigh risk
- Assess potential impact throughout lifecycle of nanomaterial

In early 2006, the Government supported the establishment of Nanotechnology and Advanced Material Incorporated (NAMI) under the direction of HKUST to facilitate nanoresearch in Hong Kong and transfer of nanotechnology to local industries. This undoubtedly elevates HKUST to a leading position in nanotechnology development in Hong Kong. It also brings into focus the need to integrate EHS assessment into nanotechnology development all the way from research, pilot production, manufacturing, consumer use and final disposal. This presents a great opportunity for us to realise the idea of "Responsible Nanotechnology", *i.e.* to not only take advantage of the benefits that nanotechnology brings us, but also ensure that we do not enjoy these benefits in the expense of health, safety or the environment.

HKUST

Nanomaterial and nanotechnology is a rapidly evolving field, and is one of the focus areas of research at HKUST. The environmental health and safety knowledge about nanomaterials and nanotechnology is also being generated by many researchers worldwide. Even though there are considerable uncertainties, it is prudent to take precautionary measures to minimize the potential of adverse health, safety, and environmental impacts. The following are elements of the Nanosafety Program at HKUST. Due to rapid development of this field, these program elements will be revised when necessary.

- *Facility Review*: HSEO has been involved in the planning of laboratory facility for the Institute of Nanomaterials and Nanotechnology (INMT). When faculty members establish other research facilities for nanomaterials and nanotechnology, they should involve HSEO at the planning stage to ensure necessary safety installations are included in the physical facility, safe operational protocols are developed, and suitable personal protective equipment is prescribed and made available to lab workers.
- *Research Proposal Review*: A nanosafety section has been added to the research proposal safety review form under the Committee of Research Practices (CRP), where principal investigators are requested to declare whether any nanomaterial will be used, and if so, the physical and chemical properties of the nanomaterial, known toxicity, use conditions, proposed control measures, and final disposal arrangement. Based on the information provided, the Safety Panel of CRP will suggest necessary precautionary measures.
- *Laboratory Visit*: At this early stage of development, one of the crucial elements is to establish a database of nanoresearch activities on campus. Based on the information collected from the research proposal review, HSEO staff will visit research groups working on nanoscience and nanotechnology to collect information on the nanomaterials, work processes, potential hazards involved, and waste disposal practices. Since safety standard is yet to be

developed, available guidelines such as those from NIOSH and HSE will form the basis of recommendations on EHS issues.

- *Safety Precautions and Safety Installations*: As mentioned in Section E above, HEPA filter is a proven method to contain and control airborne nanoparticles. Research should conduct all processes that may give rise to airborne nanoparticles inside one of the following safety devices:
  - Biological safety cabinet (BSC), if a dust-free environment is not critical, a Class 1 BSC can be used. Otherwise, a Class 2 BSC can provide both a dust-free environment and protect lab workers at the same time. However, BSC cannot contain volatile or toxic chemicals, if these are to be used together with nanoparticles. One should use a fume hood equipped with HEPA filtration
  - Fume hood equipped with HEPA filtration
  - Glove box

  Be reminded that BSC, fume hood and HEPA filter all need periodic testing to ensure they are properly functioning. Please contact either FMO-LS or HSEO if you have questions about such periodic testing.
- Safety Precautions and Personal Protective Equipment (PPE)

  Laboratory workers handling nanoparticles or other nanomaterials should follow these guidelines:
- Always ensure personal hygiene and observe good general laboratory practices, including no drinking and eating in the laboratory, always wearing lab coat and safety glasses, etc.
- Consider the hazards of the parent materials in assessing hazards of nanomaterials. However, remember the nano features of the materials most likely will present many other physicochemical and toxicological properties not found in the parent materials.
- Never handle nanomaterials with bare hands or

expose such materials to bare skin. Until further information is available, use chemical resistant gloves and protective clothing. The material of these PPE items should at least be protective against the carrier medium, such as the solvent, containing the nanomaterials.

- If it is necessary to handle nanoparticles outside of a safety installations specified in the previous section, use appropriate PPE such as a properly fitted elastomeric respirator with HEPA filter cartridge. Please contact HSEO for requirements of a respirator user. If the exposure risk to nanoparticles is considered low, particulate respirators with a filtration efficiency equivalent to HEPA (*e.g.* the 100 series approved by NIOSH or P3 by EC standard) may be used.
- A vacuum cleaner equipped with a HEPA filter may be used to remove nanomaterials from reactors or work areas. However, the integrity of the HEPA filter must be periodically tested to ensure proper functioning.
- Dispersed airborne nanoparticles, like many other airborne fine particles, can become a fire and explosion hazard. Care must be taken to prevent ignition.
- Equipment in contact with nanomaterials should be evaluated for residual hazard before servicing, reuse for other purposes, or disposal.
- Contact HSEO for assistance in evaluating or controlling hazards related to nanomaterials.
- Waste Disposal
  - Waste containing nanomaterials should be disposed as chemical wastes as a minimum. See Chapter 16 for general chemical waste disposal procedures.
  - Researchers should seek to remove the nanomaterials from the waste stream where possible.
  - Alternatively, researchers should destruct the

nano features of the materials where feasible. For examples, acid may be added to waste solution containing quantum dots to dissolve the nanoparticles before disposal.

- Following the principle of "Responsible Nanotechnology, and the concept of minimizing potential environmental impact, researchers should develop suitable destruction method for nanomaterials as a part of their research.
- Contact HSEO for assistance in disposal of nanomaterial waste.

- *Exposure Monitoring*: Even though standard method has not been established for measuring exposure to nanoparticles yet, some conventional exposure monitoring methods can be employed to provide an estimate of exposure. Some field instruments can also be used to measure airborne nanoparticles. HSEO will continue to monitor the development of exposure measuring technology for nanomaterials.
- *Continuous Improvement*: As the field is rapidly evolving, HSEO will continue to monitor the development, especially the environment, health and safety aspects, and will introduce necessary amendments to the nanosafety program. We urge all researchers to integrate EHS considerations into their research projects, and contribute their expertise to the responsible development of nanotechnology at HKUST.

The continuing revolution in computer hardware is packing ever more logic gates in ever smaller volumes. Unfortunately this produces ever more heat, which will limit the feasible packing density and performance unless the energy dissipated by each logic operation can also be dramatically reduced.

To reduce the energy dissipation of a logic operation below ln(2) kT (near thermal noise) requires the use of reversible logic for fundamental thermodynamic reasons. Extrapolation of current trends suggest this limit will become significant within one to two decades. Many real devices can be viewed as electrically controlled switches, so a method of using an abstract switch in a reversible manner is useful. Two methods of using switches to implement reversible computations are discussed in this paper.

The first method has an energy dissipation which is proportional to the square of the error in the voltage, while the second method has an energy dissipation which can in principle be reduced indefinitely by slowing the speed of computation.

The first method is basically an extension to "pass logic" which has been previously used with both nMOS (hot clock nMOS) and CMOS transmission gates to achieve low energy dissipation. The second method is a novel thermodynamically reversible logic system based on CCD-like operations which switches charge packets in a reversible fashion to achieve low energy dissipation. There is now a fairly extensive literature on reversible computation which shows that the energy dissipation per device operation cannot be reduced below ln(2) kT (where k is Boltzman's constant and T is the temperature; kT is roughly the thermal energy of a single atom) if the device is not reversible.

For the last 50 years the energy dissipation per gate operation has been declining with remarkable regularity. Extrapolation of this trend shows the energy dissipation per device operation reaching kT by the year 2015. (This assumes that T is 300 Kelvin-more on this later). To gain some perspective on this consider that an "AND" gate which has a power supply of one volt and which allows 100 electrons to go from that one volt supply to ground during the course of a switching operation will dissipate 100 electron volts. Although 100 electron volts is about 4,000 times kT (and well above the theoretical limit), it will be difficult for simple improvements of irreversible approaches to reach even this level of energy dissipation.

Extrapolating present trends, we should reach 4,000 kT before the year 2000, *e.g.*, within ten years. While reversible logic is an absolute necessity if we are to reduce energy dissipation per device operation below kT, it is a useful heuristic for designing systems that have low energy dissipation even when the actual energy dissipation is well above kT.

Even if we do develop irreversible devices that approach ln(2) kT, a computer operating at room temperature at a frequency of one gigahertz with 10^18 logic gates packed into a volume not much larger than a cubic centimeter would dissipate over three megawatts. The drive for ever greater computational power with ever more densely packed logic elements will eventually require that a single logic operation dissipate orders of magnitude energy less than kT.

If we are to realise the full potential of nanoelectronics and molecular logic devices, at some point in the future we will be forced to use thermodynamically reversible logic elements for a substantial fraction of logic operations.

We can state quite confidently that one of three things will occur:

- The historic rate of decreasing energy dissipation per device operation will slow or halt in the next one or two decades or
- We will operate computers at lower temperatures or
- We will develop new methods of computing in a reversible way that can beat the kT barrier.

The first option is unattractive. First, cooling is a major problem. The heroic cooling methods used in the Cray 3 supercomputer to remove the heat generated by the computer's operation suggest that failure to reduce energy dissipation per gate operation would be a major limiting factor in future computer performance.

While both air-cooling a single chip that dissipates 150 watts and water cooling of a chip dissipating 790 watts/cm^2 have been demonstrated, and more effective cooling methods should be feasible in the future, we will eventually reach a limit to heat removal.

Second, in many applications power is limited. Portable computers, remote sensors and other isolated systems have only a limited amount of power available. Third, although the raw cost of electrical power is not yet a major limitation it would become so in the future if reductions in energy dissipation did not keep pace with advances in other areas.

The Wall Street Journal said "Computer systems currently account for 5% of commercial electricity use in the U.S., with the potential to grow to 10% by the year 2000." Operating computers at a lower temperature will not reduce overall energy dissipation. If we operated future devices at 3 Kelvins we would reduce kT by a factor of 100 and so could reduce energy dissipation per logic operation by a similar factor-but for fundamental thermodynamic reasons the coefficient of performance of the refrigerator for the system can be at best 3K/(300K–3K) or about 0.01.

Thus, the lower energy required per gate operation will be almost exactly balanced by the increased energy needed by the refrigerator. In many applications refrigeration is not an attractive option. Lap top and portable computers, embedded systems, various "smart" appliances and other applications make the use of refrigeration undesirable. Factors other than net energy savings can make low temperature operation worthwhile. For example, some potentially attractive devices don't operate at higher temperatures.

In large computers operating in stable environments (the traditional computer centre, for example) refrigeration might be attractive, particularly if it permits the use of devices that provide much better performance but which require low temperature for their operation (Josephson junction devices might be an example). Refrigeration *per se*, however, does not seem too attractive. Finally, and most attractively, we could use reversible computation to reduce energy dissipation. This would, in principle, allow energy dissipations indefinitely below kT per logic operation.

While some barrier should eventually be encountered, the use of reversible logic should allow us to continue current trends in energy dissipation per logic operation for the longest possible time. It is interesting to note that certain uses of pass logic, transmission gates, and hot clock nMOS do in fact perform some logic operations in a thermodynamically reversible fashion.

Research in the design and utilization of reversible logic to futher lower energy dissipation will simply recognize and make explicit a trend that has already begun. Further research in this area is the most appropriate response to the rather limited range of possibilities that face us.

REVERSIBLE ARCHITECTURES

One concern about this approach is the need to use reversible computer architectures. Such architectures are entirely feasible. A wide variety of computations can be done in a logically reversible manner. Bennett concluded: "...for any e > 0, ordinary multitape Turing machines using time T and space S can be simulated by reversible ones using time O(T^(1+e)) and space O(S log T) or in linear time and space O(S T^e)." This general result shows that even arbitrary irreversible computations can be mapped into reversible computations.

Specific irreversible computations can often be mapped into reversible computations with little or no loss of efficiency. Even if we do not adopt new reversible computer architectures, simple applications of reversible computation could be made within the framework of existing architectures.

A typical computer executes a sequence of instructions, and each instruction will typically change the contents of a single register or memory location. Although loading the result of the instruction into a register will normally be irreversible (it erases the previous contents of the register) it is still the case that all other operations performed by the computer during instruction execution could in principle be made reversible.

Thus, although we would have to dissipate roughly kT energy for each bit in the output register for each instruction execution, we need not use dissipative logic throughout the computer (as is currently done).We can go one step further without making significant changes in computer architecture: the simple register-register add instruction $R_1 = R_1 + R_2$ is logically reversible. With proper hardware, this particular instruction could in principle be made to dissipate as little energy as we desired. While non-reversible instructions (*e.g.*, R1 = 0) would still dissipate greater energy, the overall energy dissipation of the computer could be reduced significantly if a reversible approach were developed and used to implement the reversible instructions. Of course, once the energy-wasting instructions were identified, compilers would learn to avoid using them. Today, compilers can optimize for speed of computation and for the amount of memory used.

In the future, compilers could also optimize a programme to minimize the energy dissipated during computation, *e.g.*, to minimize the number of bits that are erased. This would provide an entirely evolutionary path from the current irreversible designs to computer architectures that were as reversible as was practically feasible. While it is perhaps not obvious how far this trend can go it is clear that a very large percentage of computer operations can be made reversible-perhaps a remarkably large percentage.

REVERSIBLE DEVICES

Several reversible devices have been proposed over the years. Von Neumann and independently Goto proposed the parametron which encodes information in the phase of an oscillation. Fredkin and Toffoli proposed an electronic reversible logic family which used switches, capacitors and inductors. By turning the switches on and off at just the right times, charge could be transferred between the capacitors through the inductors with minimal energy loss.

By increasing the size of the inductors it would be possible to create a family of ever slower but ever more energy efficient circuits. As the LC time constant is increased, the energy dissipation can be decreased to an arbitrary extent.

A proposal by Likharev based on Josephson junctions "...is particularly significant, because it is a genuine example... of a system that has frictional forces proportional to velocity" according to Landauer. The parametric quantron is based on Josephson junctions and operates at low temperatures. It should dissipate less than kT when operating at 4 kelvins and with a switching time of 1 nanosecond. Modern "high temperature" superconductors might allow operation at relatively high temperatures, including that of liquid nitrogen.

Even these temperatures would still be a significant disadvantage. Molecular-scale mechanical reversible "rod logic" has been proposed by Drexler which should dissipate less than kT when operating at room temperature at a speed of 50 to 100 nanoseconds. Drexler's proposal is also significantly smaller than the parametric quantron. While the ability to fabricate Drexler's proposal is some decades away, it provides a good argument that molecular-scale reversible devices should be feasible.

Although the parametric quantron can be used in a reversible fashion to achieve low energy dissipation it uses magnetic fields to couple logic devices, requires low temperatures to operate and will likely prove resistant to scaling to the molecular size range. By contrast, Drexler's mechanical proposal might well prove smaller than even the smallest of future electronic devices. Mechanical devices depend fundamentally on the position of the nuclei of the atoms of which they are made. The position of the nucleus of an atom can be much more precisely known than the position of an electron: quantum uncertainty in the position of the electron is much greater because the mass of the electron is much smaller.

If device function depends on the position of an electron then device size will be larger than if device function depends on the position of the nucleus. Such molecular mechanical proposals will, however, almost certainly be slower because of the greater mass of the nuclei of atoms as compared with the mass of electrons. In this paper we discuss simple methods of using relatively conventional electronic switching devices in a reversible fashion which can be scaled to a very small size. The methods described here allow a simple voltage-controlled switch to be used in a reversible manner, both to implement combinational logic and also to iterate an arbitrary reversible logic function.

Other methods are feasible. For example, helical logic—an elegant form of reversible electronic logic—has recently been proposed by Merkle and Drexler. Voltage controlled switches of various types have long been used in electronic devices. Not all voltage controlled switches are suitable for use in low energy designs because of their intrinsic energy dissipation.

Traditional relays actuated by electromagnets dissipate comparatively large amounts of energy. Even if a switch were suited for low energy operation it would still dissipate energy if it were turned on when there was a voltage across it. The abrupt rush of current that normally occurs when a switch is turned on would insure this. Most previous logic designs based on switches operated in this manner (though the proposal by Fredkin and Toffoli avoids this problem, and the proposal by Seitz et. al. eliminates dissipative switch closures for combinational logic, although not for sequential logic).

The switch is turned on for the purpose of draining charge when the charge is no longer wanted. That is, a logic "1" might be encoded by a charge on a capacitor. Turning on a switch to drain away that charge would dissipate energy even if the switch itself were perfectly non-dissipative. A related but distinct line of research also emphasizes the importance of energy dissipation, but assumes that the losses involved in dissipatively switching a wire from one logic state to another are fixed and cannot be altered, *e.g.*, that slowing the switching speed will not reduce energy dissipation.

Even assuming that this is true, clever design of the circuit can greatly reduce the number of times it is necessary to take such a dissipative action. Further research which combines both this approach and the approaches described here would seem fruitful. Even relatively recently some authors have argued that switching devices must fundamentally dissipate more than kT regardless of how slowly they operate. It is worth noting, therefore, that charging and discharging a condenser need not dissipate the energy involved.

For example, a simple LC oscillator will repeatedly charge and discharge a condenser. Although the energy stored on the condenser is 1/2 CV^2, this energy need not be dissipated during a charge/discharge cycle.

If the Q of the circuit is high the energy dissipation per cycle will be small. In addition, the energy dissipation of such an oscillator depends on frequency. By using larger condensers and inductors the frequency of operation can be slowed and the fraction of energy dissipated per cycle reduced. If all linear dimensions of a device are scaled in size in proportion to some characteristic length L then capacitance and inductance are proportional to L while frequency and resistance are proportional to 1/L. The Q of the circuit is proportional to L so the percentage of the total energy in the oscillator that is dissipated per cycle can be reduced simply by increasing all linear dimensions equally.

Different scaling methods can produce better results. If we assume that a fixed capacitive load is charged through a fixed resistance by a voltage that is ramped up over a period of time T, then the energy dissipated by the resistance is approximately 1/2 CV^2 × 2RC/T (for RC << T).

This dissipation can be reduced as much as we might want by increasing T (and leaving the circuit components unchanged). Koller and Athas have called this adiabatic charging. We will call it "thermodynamically reversible" or simply "reversible," relying on context to make it clear that we are referring to an asymptotically non-dissipative process. Note that the term "reversible" can also be used to mean "logically reversible" with no implication that it is asymptotically non-dissipative.

A logically reversible process might be highly dissipative, so it is sometimes necessary to distinguish between "logically reversible" and "thermodynamically reversible. The situation is quite different for a step function in which the voltage is ramped up instantly (T=0). This process dissipates 1/2 CV^2. Charge falls through the fixed potential drop in a dissipative fashion much as water goes over Niagra falls. Although this is the method in common use today to charge and discharge the various capacitive loads in computer circuits, it is not the method that we will be considering.

By using reversible charging, further increases in the switching time can be used to decrease the energy dissipation per logic operation. This, of course, results in a slower speed of operation leading to a trade-off between speed and energy dissipation. How to arrange matters so that current flows smoothly and reversibly is the principle topic of the rest of the paper.

This is challenging because the basic switching devices are:

- Abrupt and discontinuous and
- Pass through a dissipative intermediate state (which is neither fully on nor fully off) while switching.

THE FIRST METHOD

The first method uses a conventional voltage controlled switch but in a somewhat unusual manner. In essence we never turn a switch on or off when there is either a voltage across it or current going through it. This mode of operation is reversible. We extend this approach to implement a sequential circuit which can iterate the computation of a reversible logic function. It is also related to "dry switching," a method for using relays which avoids turning relays on or off when there is a voltage across them. The primary purpose of dry switching is to avoid deterioration of the relay contacts, rather than minimize switching energy.

We also adopt the following convention for turning a switch on and off: when the control input to the switch is at logic "0" (low) the switch is turned on and current can flow from the input to the output or from the output to the input. When the control input is "1" (high) the switch is turned off and current cannot flow. The reader should note that this convention is not the usual one. We adopt it instead of its opposite —where a "1" turns the switch on and a "0" turns it off— because it is more intuitive when applied to the second proposal discussed later in this paper, the CCD-based reversible logic element.

The proposed mode of operation is fundamentally clocked and multi-staged. The outputs of stage i drive the inputs of stage i+1. For the moment, we assume the output of stage 0 is simply given. That is, a set of input lines are assumed to have fixed voltages upon them representing the input data.

that is required to support computation in a reversible fashion.

The single primitive operation is: Transfer charge out of a source bucket and into an empty destination bucket if a "condition" bucket holds a "0", but do not transfer the charge if the condition bucket holds a "1". If the transfer does not take place (the condition bucket is "1"), the destination bucket need not be empty. "Empty" in this context means "has no charge carriers present." That is, an "empty" bucket is in logic state "0." We will call this device a 3-bucket Reversible Charge Transfer device, or (where context makes it clear which type of RCT we are referring to) simply an RCT.

The RCT is a reversible DCCL device. At a logical level the RCT is related to (but somewhat different from) a switch gate implemented using charge packets instead of colliding billiard balls. The particular logic function described here is but one example of the range of logic functions that involve the reversible transfer of charge from one bucket to another.

During normal operation, RCT devices:

- Never merge or split charge packets
- Never discharge charge packets to ground, and
- Move charge packets in a thermodynamically reversible fashion from a set of source buckets to a set of destination buckets, conditional on the presence or absence of charge in a set of condition buckets.

In this context thermodynamic reversibility means that energy dissipation per device operation, in the absence of manufacturing defects, could be made much smaller than kT simply by slowing the speed of operation.

The number of charge packets is conserved and therefore RCT devices are conservative logic devices (in conservative logic devices, the number of 1's is conserved during the computation, so the number of 1's at the input is the same as the number of 1's at the output).

We have described the 3-bucket RCT, which has one condition bucket, one source bucket, and one destination bucket. More complex RCT's have more buckets and the pattern of charge transfer is more complex. Some of these RCT's will be considered later.

The simple 3-bucket RCT operation might be written in C as:

- If (Condition==0) {Destination = Source; Source = 0};
while a somewhat briefer notation would be:
- If Condition is 0 then Destination = Source. (Note that in this briefer notation, the "=" sign is used to indicate both the assignment of the Source value to the Destination, and setting the Source to 0 after the assignment).

The precondition that must be true prior to execution of the 3-bucket RCT is:

(Destination==0) OR (Condition==1)

Violation of this precondition would result in an unwanted dissipative step, which is banned in RCT devices. While it might at first seem that satisyfing this precondition would make the 3- bucket RCT data dependent, it is in fact possible to design a circuit in which we know that the precondition is satisfied without knowing what values are presented as inputs to the circuit. A logic diagram showing the possible initial and final states of a 3-bucket RCT is shown in Figure. On the left, the legal initial states of the RCT are shown, while on the right the final states (after the sequence of clock signals that cause the device to step through a single cycle of operations) are shown.

It is interesting to note that there is only a single legal, non-trivial change of state: when the input is 1, the condition is 0, and the output is 0. In all other cases either the input and output states are the same or the input state is illegal. The illegal states would cause undesired energy dissipation.

FREDKIN GATE WITH 3-BUCKET RCT

The 3-bucket RCT is logically complete, for we can implement a Fredkin gate with it. The construction used here is related to (though diffrent from) that used by Fredkin and Toffoli for constructing a billiard ball Fredkin gate from a billiard ball switch gate.

A Fredkin gate has three inputs and three outputs. The inputs are the control input A, and two signal inputs B and C. The outputs are A′, B′, and C′. If the control input A is 0, then

A′ = A, B′ = B, and C′ = C. If the control input A is 1, then A′ = A, B′ = C, and C′ = B. That is, the output of the Fredkin gate is identical to the input if A = 0, but B′ and C′ are exchanged if A = 1.

The following sequence of simple RCT operations will implement a Fredkin gate: A, B, and C hold arbitrary input values. InitiallyOne is set to a logic "1" (the presence of an electron packet). All other variables are initialized to logical "0" (the absence of an electron packet).

- If A is 0 then C′ = C.
- If A is 0 then B′ = B.
- If A is 0 then NotA = InitiallyOne.
- If NotA is 0 then C′ = B.
- If NotA is 0 then B′ = C.
- If AlwaysZero is 0 then A′ = A.
- If A′ is 0 then InitiallyOne = NotA.

Steps 1 and 2 simply transfer charge from B and C into B′ and C′ if A is 0. B′ and C′ are initially empty, so this transfer satisfies the RCT preconditions.

Step 3 computes the logical negation of A and leaves that logical negation in "NotA."

Steps 4 and 5 transfer charge from B and C into C′ and B′, exchanging the outputs. The precondition is satisfied, for if A is 0, then NotA will be 1 and the transfer will not take place. If the transfer does not take place, the prior contents of B′ and C′ are irrelevant.

If A is 1, then NotA will be 0 and the transfer will take place. In this case, B′ and C′ will be empty (have logic values 0) and so the transfer will take place correctly.

Step 6 copies A into A′ unconditionally.

Step 7 restores the value of InitiallyOne if it was altered during the computation.

This implementation of a Fredkin gate will take three arbitrary input values and, after a sequence of RCT operations, produce three output values. We could clearly apply N Fredkin gates to 3N input values and produce 3N output values at the same time. We require that all output values be initially 0. After finishing the sequence of clocking operations, the inputs are all 0 and the outputs hold arbitrary values. At this point, the "input" and "output" can be logically exchanged (*e.g.*, no physical operation takes place, but we relabel the "input" variables as "output," and the "output" variables as "input"). Following this logical exchange, we have reestablished the preconditions for the next sequence of operations: the (just relabeled) inputs again hold arbitrary logic values, while the (just relabeled) outputs hold logic 0's.

We now drive a (different) set of clock lines to apply the (different) Fredkin gates to generate the next "state" of the computation. Viewed globally, this is a "ping-pong" or "double buffer" scheme for computing successive values of the global state of the system. If we call the two sets of variables A and B, and we call the two logic functions for the "next state" of the system F and G, then we first compute B=F(A). B is initially all zeroes, while A holds arbitrary logic values.

On completion of the computation of F, A is all zeroes while B holds arbitrary logic values. We then compute A=G(B). Because G uses B as its input and A as its output, the condition that the outputs be all zeroes is met during the compuation of G. This returns us to the state where A holds arbitrary logic values and B holds all zeroes. We can repeat this cycle indefinitely.

IMPLEMENTATION OF A 3-BUCKET RCT

A planar layout of a 3-bucket RCT is shown in figure. Many kinds of implementations would be feasible, for example Silicon-On-Insulator technology (the FET used for the transfer switch would be "tilted on its side" in this approach). Note that many of the conductive paths are really semiconductors. Each bucket also has one semiconductor plate. The gate is also a semiconductor. Use of metallic wires for these components would compromise device function. A planar RCT is described in a later section.

This RCT can be viewed as a combination of three CCD potential wells and a FET. A charge packet is first transferred from the "condition" potential well to the gate of the FET. The charge packet in the "source" potential well is then moved to the source of the FET and (the charge on the gate of the FET permitting) through the channel and out the drain of the FET into the "destination" potential well.

Tompsett proposed sensing the presence or absence of a charge in a CCD potential well by connecting the potential well via a doped region connected ohmically to a metalization that became the gate electrode of a FET. Charge injection from the gate into the doped region, while small if the capacitance of the gate is small, still introduces an unwanted source of energy dissipation. Other aspects of Tompsett's proposal were also irreversible or highly dissipative. In the present proposal, we require that both the gate of the FET and the connecting path from the potential well to the gate be a semiconductor. The only charge carriers in this semiconductor path would be those deliberately introduced as a charge packet.

An alternative approach which might be more convenient in a simple planar implementation would be to introduce a capacitive coupling into the path from the potential well to the gate. The gate could then be metallic, one plate of the capacitor (the plate connected to the gate) could also be metallic, but the other plate and the connection to the potential well could be a semiconductor.

Potential could be coupled through the capacitor, but charge carriers would be prevented from moving into or out of the semiconductive region and hence charge injection from the gate electrode would not be a problem. Initially, a charge of unknown value is present in the source bucket. The destination bucket is either empty (if the condition bucket holds a 0) or has an unknown charge (if the condition bucket holds a 1). Clocks 1, 2, and 3 are all positive. All charge carriers (electrons) have been drawn onto the semiconductor plates of the buckets.

The conventional plates are connected to the clocks. There is a net positive charge on each bucket creating a potential well which holds the electrons in the bucket and prevents them from migrating away. The channel regions of all switches have no charge carriers present. Clock 4 is driven negative with a large voltage. This large voltage cuts off and isolates the three buckets involved from any other buckets to which they might normally be connected. Additional switches controlled by clock 4 might be required to further isolate the three buckets in a larger circuit, but are not shown in this small example.

The two gray channels do not actively participate in the process of reversible charge transfer, they are present merely to show that connections to other circuit components will be present in a real system. Clock 3 is driven negative, forcing charge from the condition bucket onto the gate of the transfer switch. If no charge is present on the condition bucket, no charge will be put onto the transfer gate, while if charge is present then it will be transferred to the transfer gate, cutting off the transfer switch. Note that the circuit is so arranged that only a medium voltage will be developed on the transfer gate, and this medium voltage can be contained by the high voltage of the cut off clock, clock 4.

Clock 2 is driven negative. If there is charge on the destination bucket, then the transfer switch is turned off and the charge on the destination bucket goes nowhere. If there is no charge on the destination bucket, then whether the transfer switch is turned off or on is immaterial, for no charge will be transferred.

Clock 1 is driven negative. If the transfer switch is turned off, no charge will be transferred. If the transfer switch is turned on, charge will move gradually through the transfer switch into the destination bucket. Because clock 2 is negative, charge will not "fall down hill" in a dissipative fashion when moving into the destination bucket.

Clock 1 has a low voltage so a low voltage is generated during the transfer operation. The medium voltage on the transfer gate is sufficient to block charge transfer if charge transfer is not supposed to take place.

Clock 2 is driven positive, allowing charge to flow smoothly into the destination bucket. When clock 2 reaches its full positive value, all the charge is held on the destination bucket and no charge carriers are present in either the source bucket or the transfer switch channel region (if the transfer switch was turned on).

Clock 1 is driven positive, which has no effect if the transfer took place and simply allows the charge to return to its original state if the transfer did not take place.

Clock 3 is driven positive, allowing the charge on the transfer gate to return to the condition bucket.

Clock 4 is driven positive, opening the cut off switches that isolated the three buckets from outside influence during the reversible transfer operation.

This cycle of operations reversibly transfers charge from the source bucket to the destination bucket if the condition bucket holds a logic "0" (*e.g.*, no charge).

Energy Loss

There are losses during the RCT cycle of operations caused by current flowing through a resistive medium. The power lost will be, Power Loss = I^2 R where I is the current and R the resistance. Current is proportional to the frequency of operation of the circuit, so if the frequency is low the current will also be low. Therefore, power losses from resistive losses will fall off with the square of the frequency: Power Loss is proportional to f^2. Because the number of operations per second is also lower at a lower frequency, the actual energy loss per device operation is proportional to the frequency of operation: Energy Loss is proportional to f. By reducing the frequency of operation, the energy loss per operation can be reduced to whatever extent is desired. It should be remembered that "low frequency" operation is relative: CCD's with 0.9997 charge transfer efficiency operating at 1 gigahertz have been demonstrated.

More specifically, the electron drift velocity vd equals the electric field E times the mobility m:

$$vd = Em$$

If we let d be the distance traveled by an electron as it moves from one bucket to the next, and t be the time of a single logic operation, then it is approximately the case that:

$$d = vd\ t$$

The total energy dissipated by the movement of n electrons as they move from the source to the destination is simply the force times the distance times the number of electrons, or:

$$E_{dissipated} - neEd$$

where Edissipated is the energy dissipated, n is the number of electrons in a packet, and e is the charge of an electron.

Putting these together yields:

$$E_{dissipated} = ned^2/mt$$

If we assume that the distance traveled by the charge packet as it moves from one bucket to the next is.1 microns (10^{-5} centimeters), that one packet has 100 electrons, that the mobility is 1800 cm^2/Vs (the approximate mobility of diamond or silicon near room temperature), that the time allotted for one operation is 1 nanosecond, and the charge e on an electron is 1.6×10^{-19} coulombs, then the energy dissipated is approximately 10^{-21} joules.

This estimate is based on a classical model and involves some significant simplifications. Most significantly, the mean free path will of necessity be less than.1 micron in a device whose maximum dimension is.1 micron, and so the meaning of the electron mobility m is somewhat obscured.

A more detailed quantum analysis is essential as the size becomes smaller. A rough approximation to such an analysis can be based on the observation that the resistance of the channel of a sufficiently small FET (*e.g.*, one in which the channel width is perhaps a few nanometers and only a single transmission mode exists) is roughly half of h/e^2, where h is Planck's constant.

This is a fundamental unit of resistance of about 26K ohms. A packet of n electrons flowing in time t produces a current of ne/t. The power produced by such a current flow is $I^2 \times R$, while the energy dissipation per logic operation is $I^2 \times R \times t$, or $(ne/t)2 \times h/(2e^2) \times t$ which simplifies to:

$$E_{dissipated} = hn^2/2t$$

For a switching time t of 1 nanosecond and a packet size n of 100 electrons, this is 3×10^{-21} joules for each packet that is passed through the channel of the FET. Under the specific conditions considered here this second approximation agrees at least in order of magnitude with the estimate based on a classical model. It is interesting to note that the two equations have significant differences. The resistance of the channel in the classical case varies with the length of the channel. The length of the channel is less significant in the nonclassical case. The size of the charge packet, n, also influences energy dissipation in different ways in the two cases. In the classical case, the energy dissipation is a function of n rather than n^2.

In the classical case, two electrons drifting under the influence of the same electric field in separate regions of the channel will dissipate twice as much energy as a single electron. In effect, the resistance has been halved because the number of charge carriers has been doubled.

A similar effect would occur in a doped semiconductor if the doping density were doubled. If we assume a narrow channel that can only support a single transmission mode, however, then doubling the size of the charge packet will cause more electrons to move through the same channel in the same time period, with a resulting increase in interactions among the electrons. The reader should note that no claim of universality or high accuracy is implied for either of the above equations. They are rough approximations that apply only under a limited range of conditions.

These conditions need not, in general, be true. In particular, it would seem attractive to design devices in which single electrons are always close to their ground state both during charge transport and during switching operations.

This would occur if the electron were always confined in a time varying potential well of relatively small dimensions. Devices with this property should be feasible when we are able to manufacture devices of sufficiently small size. This should have a favourable influence on energy dissipation. The major factor ignored in this analysis is the energy dissipated by the clock signals. Many methods of providing periodic clocking pulses are possible. Three methods are discussed in the next section.

The mobility can be substantially increased by reducing the temperature. Mobilities greater than 1,000,000 have been demonstrated in GaAs-AlGaAs HFET's at around 10 K. While the impact of this is unclear when the device size is much smaller than the mean free path (which occurs with such high mobilities), it does mean that larger devices could be made which would have very low energy dissipation. The energy dissipation for this type of reversible electronic logic (as estimated here) is greater than the dissipation of Drexler's rod logic operated at a cycle time of 100 picoseconds (some 10 times faster than the 1 nanosecond considered here for the RCT). When RCT device operation using single electrons becomes feasible, then energy dissipation using RCT's would presumably be reduced.

Likharev's parametric quantron has an energy dissipation which (when adjusted for the temperature difference and the differing speeds of operation) is not as good as rod logic. The parametric quantron also requires low temperature.

Rod logic should work at room temperature. It is possible that mechanical molecular logic might have very favourable energy dissipation properties at low temperature. While it is premature to draw conclusions it is not obvious that electronic devices must necessarily prove better than molecular mechanical devices in terms of energy dissipation at a given speed and temperature of operation.

CLOCKING

To insure that overall energy dissipation is reduced we need a method of providing suitable clocks that does not itself dissipate too much energy. We first consider a conceptually simple method of providing arbitrarily complex clock signals which can be made to dissipate as little energy as desired.

The loads presented to the clock in the proposals discussed here are purely capacitive. If we assume, for the moment, that the energy dissipated when a clock signal charges and discharges such purely capacitive loads is $1/2\ CV^2$, then the only method of reducing the energy dissipation is to reduce the voltage. Assume, for the moment, that we want a clock signal to increase from 0 volts to 5 volts. If we simply connect the clock line to a 5 volt source, then the energy dissipated will be $1/2\ CV^2 = 12.5\ C$ joules.

If we have 10 voltage supplies, with voltages of 0.5, 1.0, 1.5, 2.0, 2.5, 3.0, 3.5, 4.0, 4.5, and 5.0 volts, then we can successively connect our clock line to each voltage supply in turn. The voltage to the clock line would then be stepped up in 0.5 volt increments, and so the energy dissipated by a single step would be 1/2 C(V/10)2 = 0.125 C (*e.g.*, the energy dissipation per step is decreased by a factor of 100 because the size of the voltage step is decreased by a factor of 10).

To increase the voltage from 0 to 5 volts involves 10 steps, and so we will dissipate 10 × (0.125 C) or 1.25 C joules. This is 10 times smaller than the 12.5 C joules we would have dissipated using a single 5 volt supply. Thus, be using 10 DC power supplies and switching the clock line successively from one power supply to the next, we have reduced energy dissipation by a factor of 10. Clearly, we can reduce energy dissipation by a factor of N if we use N DC power supplies.

Notice that the larger we make N the more time we must take to increase the voltage from 0 to 5 volts. Each time we take a step in voltage we must wait somewhat longer than one RC time constant for the voltage to settle at the new value. This settling time will remain about the same for each step, even as we decrease the size of the voltage step by increasing N. As N increases and the number of steps increases, the total time taken for N steps will also increase.

This approach is very flexible because the clock waveform can rise and fall at arbitrary times. Complex clock signals can be easily generated. There are many alternative clocking schemes. While the method just described is asymptotically nondissipative if a sufficiently large number of DC power supplies are used, it still involves moving electrons through wires to carry the clock signal throughout the circuit.

This does not appear to be the major limiting factor in energy dissipation when using today's planar technology. However, if in the future we use single electron logic devices, then moving large numbers of electrons over long distances to provide the clock signals would seem inappropriate.

There are approaches to clock distribution which avoid the need to move charge through wires. A conceptually simple approach is to provide clock distribution by the use of charges that are fixed to a rotating tube or disk. A facing disk or a smaller concentric drum would carry the logic circuitry.

As the disk spun the charges fixed to its surface would move, but this motion would not involve electrons flowing within a conductor.

The electric charges on the surface of the disk would then move past the logic elements on the facing surface of the opposing disk, providing the repetitive voltage changes required to clock the logic circuitry. In principle this method of moving charges would result in no conventional resistive losses at all because the charge would not be moving through a conductor.

There would still be radiative losses caused by the acceleration of the charged particles, and electrostatic attraction and repulsion would produce forces between the charges on the two disks which would lead to losses simply by alternately compressing and decompressing the material of the disk. While this approach might not be attractive with current technology, it might prove attractive at some point in the future if strong materials could be rotated at high speeds, thus providing a high interfacial velocity between the two opposing surfaces; and the pattern of charges on the surface could be both precisely controlled and have high resolution.

This should eventually be feasible, though whether or not it will prove competitive with other approaches is uncertain. Another method of providing a time-varying electric field would be to use a simple rotating electric field. Such a rotating electric field could then be used to clock the circuit. While the methods described here do not obviously lend themselves to such an approach, helical logic uses such rotating electric fields to drive thermodynamically reversible logic operations.

Charge is transported along a helix much as water is transported along an Archimedes screw. Single electron versions of helical logic should eventually be feasible. Rotating electric fields can be provided by several methods, including cavity resonators or multiple LC oscillators connected to plates arranged in an appropriate pattern in space.

OTHER RCT'S

The 3-bucket RCT is not the only type of RCT that is useful. For example, if we use two condition buckets C_1 and C_2, and require that the charge packets in C_1 and C_2 be complementary (*e.g.*, C_1 is the negation of C_2); and if we use a single source bucket S and two destination buckets D_1 and D_2; then we can implement the primitive "Transfer charge from S to D_1 if C_1 is false, but from S to D_2 if C_1 is true." We will again need a pre-condition that effectively prevents the merging of two packets; we omit the precise statement of this precondition.

This 5-bucket primitive is logically complete (we leave this as an exercise for the reader) and has certain implementation advantages.

In particular, because both C_1 and its complement C_2 are available, we need only "steer" the charge in the source bucket to either of the destination buckets. We need not "bottle up" the charge on the source bucket in the event the transfer is not to take place (as occurs with the 3-bucket RCT). Because the charge packet in bucket S is guaranteed to move to some destination bucket (either D_1 or D_2), the actual voltage that must be generated to prevent the charge from moving in the wrong direction is smaller than the voltage required in the simple RCT to prevent the charge from transferring to its single destination bucket.

The same logic function could also be implemented in a somewhat different fashion by using a 4- bucket RCT with a single source bucket S, a single condition bucket C and two destination buckets D_1 and D_2. This is illustrated in Figure. In this alternative implementation, the charge in C would control a single FET between S and D_1, while S and D_2 would always be connected. The key to the conditional transfer is the timing of the clock lines for the two destinations D_1 and D_2.

In particular, we would transfer charge to D_1 through the FET first. During this transfer, D_2 would be maintained in an unreceptive state (the clock for D_2 would be negative). After transfer of charge to D_1 through the FET, the clock for D_2 would be made positive.

This shift would cause charge to transfer from S to D_2 only if the charge had not previously been transferred to D_1 through the FET. Thus, charge would be moved conditionally to either D_1 or D_2 depending on the condition C. In some sense, this is logically equivalent to the simple 3-bucket RCT operation followed by the transfer of the charge in S to the second destination bucket D_2.

A 4-bucket RCT with a single source bucket S, a single destination bucket D and two condition buckets C_1 and C_2 could be used to transfer charge only if both C_1 and C_2 were at logic 0. Arbitrary switching functions could be imposed between the source and destination buckets, and the number of condition buckets could be increased arbitrarily.

A Planar Version of RCT Logic

It would be advantageous to have a planar layout for an RCT that used only sine waves for clocking. This would be easy to implement in current planar technology, and the clock signals could be generated by LC oscillators. These constraints can be met. In particular, we start by considering a 3-phase CCD as illustrated in figure. The three clocks, F0, F120 and F240, represent three sine waves offset by 0, 120, and 240 degrees respectively. Charge packets will move from left to right as the three clock signals vary sinusoidally over time. The illustration shows a top view, the black squares being the metalization regions.

The metalization typically would be on top of SiO2, with the bulk Si beneath that. A profile view of charge transfer is given in figure. To provide a switching operation, we must provide some sort of choice for the charge packets moving along the CCD. This is illustrated in figure, where the source CCD shift register divides into two offspring.

To control the direction in which the charge packets are moved, we assume that two "condition" CCD shift registers provide the needed condition packets that will cause charge to select one offspring shift register over the other. Note that this device is just a 5-bucket RCT with a somewhat different geometry and clocking sequence.

It uses 3-phase sine wave clocks to move charge through the device. An additional fourth phase (F180) is required to control the conditional charge transfer from the single large source well (the large rectangle in the figure) to the two destination wells directly to its right. In this particular planar RCT, Condition1 and Condition2 are two CCD shift registers which hold complementary charges. If a potential well in the Condition1 shift register contains a charge packet, then the corresponding potential well in the Condition2 shift register does not (and vice versa).

The condition charge packet moves from left to right. When it reaches the well clocked by F180, the presence or absence of the charge packet changes the voltage on the corresponding gate electrode. There are two gate electrodes controlled by the two complementary charge packets. The gate electrodes are between the source potential well (the large rectangular metalization) and the two destination potential wells to its right. The gate electrodes determine the destination well into which the charge packet in the source potential well moves.

The voltage of the fourth phase, F180, is chosen so that the potential on the gate electrode is somewhat lower than the potential in the two adjacent potential wells when those two potentials are equal (*e.g.*, when F120 = F240)and when the condition charge packet is absent. Thus, when the condition charge packet is absent, the charge packet in the source potential well will flow through the corresponding gate potential well and then into the corresponding destination potential well. When the condition charge packet is present, then the corresponding gate potential will be increased and the charge packet in the source potential well will be prevented from moving into the corresponding destination potential well.

In this case the charge packet in the source potential well will find there is no barrier preventing it from moving into the other destination potential well. Some clock signals are marked with a prime, as F′. The clock lines marked with a prime are the same as the those that are not so marked except for their DC bias level. Because the condition shift registers and the source/destination shift registers share the metalization for the F180 clock, and because basically different things must occur when a charge packet passes beneath the F180 metalization in the two cases, the DC bias levels for the two sets of clock signals need to be separately adjustable.

Otherwise, when the gate electrode blocked a charge packet from moving from the source to the unselected destination, the same voltages applied to the metalizations on the corresponding condition shift register would likewise prevent the condition charge packet from moving along the condition shift register: this, of course, would cause problems. By adjusting the DC bias on the condition register clocks, we can insure that condition charge packets will move from left to right in a regular manner regardless of the voltage variation on the F180 clock caused by the presence or absence of the condition charge packet.

In the same way, by adjusting the bias on the source/destination clocks, we can insure that the voltage variation on the gate electrodes will prevent charge packets from moving into the wrong destination, thus insuring reliable switching action. In a real circuit, many RCT devices would be interconnected by a complex pattern of CCD shift registers. In most cases, this will force the data in one CCD shift register to cross over other shift registers. The use of a planar geometry makes the design of such a cross-over nonobvious.

A relatively simple approach would be to use two crossing sets of metalizations which define two CCD shift registers, and then apply a bias voltage to inactivate one of the shift registers.

A charge packet would be carried along the active shift register and would not be diverted into the inactive shift register because the bias voltage applied to the inactive shift register would eliminate the potential wells into which the charge packet might fall. By alternately inactivating one or the other shift register, charge packets could be carried across in first one and then the other direction.

This can be likened to a traffic crossing with a traffic light, where two cars approaching at right angles cannot simultaneously occupy the intersection. The traffic light is first

green in one direction, allowing one car to cross; and is then green in the other direction, allowing the other car to cross.

To summarize this planar RCT: a charge packet moving along a CCD shift register will be moved into one of two "offspring" or "descendant" shift registers. Which descendant is selected will depend on which of two condition shift registers is occupied by a charge packet. This operation is logically universal: it can be used to implement a Fredkin gate and therefore any reversible logic circuit. Just as this planar RCT is conceptually similar to a 5-bucket RCT, so too a planar version similar to the 4-bucket RCT would also be feasible and would have the practical advantage that only a single condition would be required to operate it.

The planar 5-bucket RCT suggested here is intended to show that planar RCT's using simple sine waves for clocking are feasible. The design of an actual device might well differ significantly.

NANOROBOT DEXTERITY AND MOBILITY

Manipulation and mobility are crucial basic capabilities in most classes of medical nanodevices. Manipulation includes handling fluids, biological objects such as tissue matrix fibres or cellular elements, and nanomachines or their components. Physicians must be able to direct tissue or cell-repair nanorobots to travel to a specific site where treatment is required, and once there, to manipulate the local environment to achieve the desired results. Nanodevice mobility in vivo makes possible the rapid reconfiguration of nanomedical communication, navigation, and power systems, while ex vivo mobility allows the design of more robust diagnostic and personal defensive systems. In vivo locomotion also permits precise mapping of the internal regions of the human body across many size and time scales, both for diagnostic and for therapeutic purposes.

Adhesion and Fluid Transport

An understanding of surface forces is an essential preliminary to any study of mechanisms of manipulation or locomotion. Such adhesive forces become important at the submicron scale where nanorobotic parts are meshing with and sliding against one another. Surface adhesion forces may cause tools, parts or workpieces to stick together. Surface forces may cause airborne or solvent-immersed nanodevices to adhere to container walls, to other dry or humid surfaces, or to each other possibly causing clumping.

For a 1-micron object, the adhesive forces may exceed gravitational and inertial forces by a factor of 10^6 or more. Capillary forces also are important in fluid transfers inside nanodevices, in fluid transfers between nanodevices or between nanodevices and their operating environment, and during nanodevice tasks requiring locomotion through, or limb/object manipulation in, a watery environment. Surface tension forces are especially relevant during locomotion missions requiring passage through air/water interfaces, as might occur in the lungs. Bowling classifies adhesive forces between objects into three categories.

The first category includes long-range attractive interactions that may bring a particle to a surface and establish the adhesion contact area, such as van der Waals forces, electrostatic forces, and magnetic forces. A second category of forces are the interfacial reactions that help to define the adhesion area, most importantly the capillary forces arising from the establishment of liquid or solid bridges between particle and surface, but also including other effects such as sintering (diffusion and condensation), diffusive mixing, mutual dissolution and surface alloying, which will not be considered further here.

In the third category are very short-range interactions that may strengthen adhesion after an adhesive contact area has already formed; such forces include chemical bonds of all types and various noncovalent bonds such as hydrogen bonds that have already been described.

NANOMANIPULATORS

Numerous simple actuators have been developed for microelectromechanical systems or MEMS. A few of these designs might possibly be useful in early nanomedical systems. All major components of an articulated micromanipulator having multiple degrees of freedom, workspaces on the order of 1 mm^3, positional resolution up to ~10 nm, operating frequencies up to 10 KHz, and slewing speeds up to 4 mm/sec have been demonstrated experimentally; complete systems have even been proposed. Actuation and manipulation may be very broadly defined. For example, self-assembling and self-disassembling ~300 micron gold foil cubes have been fabricated and cycled repeatedly at ~1 Hz; "silicon origami" has also been described. Nonmechanical actuation has also been explored.

In the Magnetic Stereotaxis System, a helmet with a cubic array of six superconducting coils is used to apply a 0.2 N force to a 3.2-mm wide, 4.7-mm long permanently magnetized cylindrical pellet that is embedded in brain material, causing stepped movement of the pellet through the brain material to a specified location with ~1 mm placement accuracy without any direct physical contact.

However, in the usual definition of a manipulator, an ergomechanical transducer converts chemical, electrical, mechanical, acoustical, or some other form of energy into the mechanical energy of a manipulatory device which then exerts useful forces on objects or materials in the environment. In most cases a force-transmitting physical structure is required which may include both heavy load-bearing elements such as girders, struts, pneumatic tubes or rotary joints, and fine control elements such as stress cables, valves or switches.

An end-effector (*e.g.*, gripper, cutter or molecular jig) is usually employed to focus and redirect the application of force at the distal end of the manipulation device, or to perform specialized tasks. Sensors permit feedback control of manipulator motions as well as verification that the assigned manipulatory task has been properly executed.

Large numbers of manipulators can form ciliary arrays, allowing convenient mass transport of materials; other classes of manipulators may be useful in bulk disassembly of materials. The diverse concepts outlined in this Section demonstrate the enormous range of possible nanomanipulator designs. These concepts are not intended as specific engineering proposals but rather as illustrations of certain useful classes of manipulators representing alternative design pathways that could provide the desired capability.

No effort has been made to produce optimal systems that minimize energy dissipation, maximize speed, minimize parts count, or simplify manufacture.

BIOLOGICAL CILIA

Biological cilia are motile, hairlike cylindrical processes typically 200–300 nm in diameter, 2–20 microns in length, and ~10^{-16}–10^{-15} kg in mass. They are found in large numbers on some cell surfaces. Unicellular organisms use cilia both for locomotion and for food collection, while multicellular organisms use cilia primarily for mass transport of the environment past the cell. Each cilium is bounded by an extension of the cellular plasma membrane, rising from a cytoskeletally-anchored centriole-like basal body consisting of an array of 9 triplets of common-walled microtubules ~150 nm in diameter and 300–500 nm in length. Sprouting upward from the basal body is the primary ciliary structure, called the axoneme.

The axoneme is a cylinder of microtubules with 9 outer doublets of microtubules and two additional complete microtubules in the centre, often called the central pair. The 9 outer doublets are extensions of two of the three microtubules comprising each of the nine basal body triplets. Each outer doublet of the axoneme consists of the A tubule (a complete 25-nm outer-diameter microtubule with the usual 13 protofilaments made of tubulin dimers), the B tubule (an incomplete microtubule with only 10–11 protofilaments), and a common wall containing the protein tektin (an intermediate-filament-like protein with high tensile strength).

5

Nanotechnology in Drug Metabolism

INTRODUCTION

The vast majority of drugs are small molecules designed to bind, interact, and modulate the activity of specific biological receptors. Receptors are proteins that bind and interact with other molecules to perform the numerous functions required for the maintenance of life. They include an immense array of cell-surface receptors (hormone receptors, cell-signaling receptors, neurotransmitter receptors, etc.), enzymes, and other functional proteins.

Due to genetic abnormalities, physiologic stressors, or some combination thereof, the function of specific receptors and enzymes may become altered to the point that our well-being is diminished. These alterations may manifest as minor physical symptoms, as in the case of a runny nose due to allergies, or as life threatening and debilitating events, such as sepsis or depression. The role of drugs is to correct the functioning of these receptors to remedy the resulting medical condition.

As an example, the highest grossing drug was Prilosec, which earned 4.102 billion in sales. Prilosec is used to treat stomach ulcers and acid reflux disease. Prilosec targets a specific enzyme, the proton pump, which is located in the acid producing cells lining the stomach wall. This enzyme is responsible for the production of stomach acid. Due to genetic reasons, such as deficient enzymes that regulate acid secretion, or physiologic causes, such as stress, too much acid may be produced. This leads to ulceration of the stomach lining or acid reflux disease and heartburn. Prilosec binds to the proton pump and shuts it down, thereby diminishing the production of stomach acid and its associated symptoms.

BIOCHEMISTRY OF DRUGS

Enzymes are a subset of receptor-like proteins that are directly responsible for catalyzing the biochemical reactions that sustain life. For example, digestive enzymes act to break down the nutrients of our diet. DNA polymerase and related enzymes are crucial for cell division and replication. Enzymes are genetically programmed to be absolutely specific for their appropriate molecular targets. Any errors could have grave consequences. One can imagine the end result should blood clotting enzymes start activating throughout the body. Or consider the problems that arise when our immune system attacks our own tissues.

Enzymes ensure the specificity of their targets by forming a molecular environment that excludes interactions with inappropriate molecules. The analogy most often mentioned is that of a lock and key. The enzyme is a molecular lock, which contains a keyhole that exhibits a very specific and consistent size and shape. This molecular keyhole is termed the*active site* of the enzyme and allows interaction with only the appropriate molecular targets. Just as a typical lock is much bigger than the keyhole, the receptor is usually much larger than the active site. The receptor, as specified by our DNA, is a folded protein whose major purpose is to form and maintain the size and shape of the active site.

The most important concept in drug design is to understand the methods by which the active site of a receptor selectively restricts the binding of inappropriate structures. Any potential molecule that can bind to a receptor is called a *ligand*. In order for a ligand to bind, it must contain a specific combination of atoms that presents the correct size, shape, and charge composition in order to bind and interact with the receptor. In essence, the ligand must possess the molecular key that binds the receptor lock. This is termed *steric complementarity*. As is the case with an actual key, if a different molecule varies by even a single atom in the wrong place, it may not fit properly, and will most likely not interact with the receptor. However, the more closely the fit between the ligand and receptor, the more tightly the interaction becomes. Again, keep in mind that this is only a two-dimensional schematic. Both ligand and active site are volumes with complex three-dimensional shape.

In addition to steric complementarity, electrostatic interactions influence ligand binding. Charged receptor atoms often surround the active site, imparting a localized charge is specific regions of the active site. From physics, we know that opposite charges attract while similar charges repel. *Electrostatic complementarity* further restricts the binding of inappropriate molecules since the ligand must contain correctly placed complementary charged atoms for interaction to occur.

The main driving force for ligand and receptor binding is *hydrophobic interaction*. Nearly two-thirds of the body is water, and this aqueous milieu surrounds all our cells. In order for ligand and receptor to interact, there must be a driving force that compels the ligand to leave the water and bind to the receptor. The hydrophobicity of a ligand is what causes this. Hydrophobicity stands for 'water fearing' and is a measure of how 'greasy' a compound is. It can be roughly approximated by the percentage of hydrogen and carbon in the molecule.

This force is easily demonstrated by placing a few drops of oil in a cup of water. The oil is composed of hydrocarbon chains and is highly hydrophobic. The oil droplets will instantly coalesce into a single globule in order to avoid the water, which is highly polar. Depending upon the size of the active site, there may be a myriad steric, electrostatic, and hydrophobic contacts. However, some are more important than others. The specific interactions that are crucial for ligand recognition and binding by the receptor are termed the *pharmacophore*. Usually, these are the interactions that directly factor into the structural integrity of a receptor or are involved in the mechanism of its action. Using our lock and key analogy from above, we can imagine a lock having numerous tumblers.

There may be many keys that can sterically complement the lock and fit within the keyhole. However, all but the correct key will displace the wrong tumblers, leading to a sub-optimal interaction with the lock. Only the correct key, which presents the pharmacophore to the receptor, contacts the appropriate tumblers and properly interacts with the lock to open it. This is crucial to the design of pharmaceuticals since any successful drug must incorporate the appropriate chemical structures and present the pharmacophore to the receptor.

In the upper left frame of this figure, we see our native ligand bound within the active site. Assume that through biochemical investigation, we determine that the phenyl ring and the carboxylic acid group are vital to receptor interaction. Thus, we deduce that these two groups must be the pharmacophore that a ligand must present to the receptor for binding. In future drugs that we develop to mimic the native ligand, we must include these two pharmacophoric elements for successful binding to occur.

This is shown in the upper right derivative compound where a bicyclic group has been substituted. Because it maintains the pharmacophore and retains its complementary size and shape, it has a reasonable chance of successfully binding. However, any drug that we develop which lacks a complete pharmacophore may not interact withz the receptor target.

When a medical condition exists where a drug could be beneficial, extensive scientific study must first be done in order to determine the biological and biochemical problems that underlie the disease process. This often takes years of study in order to characterize the targets for a potential drug. The reason is that nearly all biological processes in the human body are tightly interconnected. Altering the behaviour of select receptors or enzymes may have detrimental effects with other systems.

These are the side effects that occur with nearly all drugs. Furthermore, the human body is a homeostatic machine, and always attempts to achieve equilibrium. As a result, the body will attempt to counteract any pharmacotherapeutic intervention. Once a receptor target has been established and well characterized, the process of ligand design begins. Obviously, the first consideration is that the designed ligand must complement the active site of the receptor target. Steric, electrostatic, and hydrophobic complementarity must be established as we discussed above. The pharmacophore must be presented to the receptor in order for recognition and binding to occur.

In addition to adequately binding the receptor, the biochemical mechanism of the receptor target must be taken into consideration. If the peptide bond is present at a specific position in the active site when the ligand binds, it is cleaved by the protease with the addition of water (H_2O) to form two separate fragments. If our goal is to inactivate this protease, any designed ligand cannot possess this peptide bond at the same position.

Otherwise, it will simply be cleaved by the protease, and the protease will continue to function unperturbed. However, the ligand can be modified so that the peptide bond is no longer present as shown in the bottom portion of the figure. If the enzyme then binds this ligand, the enzyme will not be able to cleave it. As such, the enzyme would be inactivated, as the ligand remains lodged in the active site.

Having characterized the active site region and the mechanism of action of the target receptor, the challenge then becomes one of designing a suitable ligand. This is, by far, the most daunting task of the entire drug design process. The optimal combination of atoms and functional groups to complement the receptor is often the natural ligand of the receptor. Unfortunately, this is usually an unacceptable candidate for a drug. This is because the natural ligand either fails to inactivate the receptor, as described above, or it is a natural substance that cannot be patented. Patent considerations are often paramount, as legal protection for the developed drug affords the opportunity to recoup the financial costs of development. Therefore, alternate combinations of chemical structures must be devised.

The design of novel ligands is often restricted by what chemists are physically able to synthesize. It is of no use to design the ultimate drug if it cannot be manufactured. The laws of chemistry dictate that each atom type has a specific size, charge, and geometry with respect to the number and types of neighbouring atoms that it can be joined to. The entire field of chemistry is predicated on the establishment of synthetic rules for the construction and manipulation of various combinations of atoms and functional groups. It is the expertise in these chemical rules that govern the ability of the synthetic chemist to design and synthesize postulated ligand candidates. Within these rules, the drug developer must creatively propose suitable chemical structures that satisfy the requirements discussed above.

Finally, there are biological considerations to the development of new drugs. The liver is the major organ of detoxification in the human body. Any drug that is taken undergoes a number of chemical reactions in the liver as the body attempts to neutralize foreign substances. This set of reactions is well characterized, and a great deal of knowledge exists as to how drugs are modified as the body eliminates them. More importantly, various chemical structures are highly toxic to biological systems, and these are also well characterized. These constraints must also be taken under consideration as novel drugs are developed.

NANOTECHNOLOGY IN DRUG DISCOVERY

The result is the determination of specific receptor targets that must be modulated to alter their activity in some way. Once these targets have been identified, the goal is then to find compounds that will interact with the receptors in some fashion. At this initial stage of drug development, it does not matter what effect the compounds have on the targets. We simply wish to find anything that binds to the receptor in any fashion.

The first step is to determine an *assay* for the receptor. An assay is a chemical or biological test that turns positive when a suitable binding agent interacts with the receptor. Usually, this test is some form of colorimetric assay, in which an indicator turns a specific colour when complementary ligands are present.

This assay is then used in *mass screening*, which is a technique whereby hundreds of thousands of compounds can be tested in a matter of days to weeks. A pharmaceutical company will first screen their entire corporate database of known compounds. The reason is that if a successful match is found, the database compound is usually very well characterized. Furthermore, synthetic methods will be known for this compound, and patent protection is often present. This enables the company to rapidly prototype a candidate ligand whose chemistry is well known and within the intellectual property of the company.

If a successful match is found, the initial hit is called a *lead compound*. The lead compound is usually a weakly binding ligand with minimal receptor activity. The binding of this structure to the receptor is then studied to determine the interactions that foster the ligand-receptor association. If the receptor is water soluble, there is a chance that *x-ray crystallographic analysis* can be employed to determine the three-dimensional structure of the ligand bound to the receptor at the atomic level.

This is a very powerful tool for it allows scientists to directly visualize a snapshot of the individual atoms of the ligand as they reside within the receptor. This snapshot is referred to as a *crystal structurc* of the ligand-receptor complex. Unfortunately, not all complexes can be analysed in this manner.

However, if a crystal structure can be determined, a strategy can then be developed based upon this characterization to improve and optimize the binding of the lead compound. From this point onward, a cycle of iterative chemical refinement and testing continues until a drug is developed that undergoes clinical trials. The techniques most often used to refine drugs are *combinatorial chemistry* and *structure based design.*

Combinatorial chemistry is a very powerful technique that chemists can employ to aid in the refinement of the lead compound. Combinatorial chemistry is a synthetic tool that enables chemists to rapidly generate thousands of lead compound derivatives for testing. As shown above in Figure, a scaffold is employed that contains a portion of the ligand that remains constant.

Subsite groups (shown in red, green, and blue) are potential sites for derivatization. These subsites are then reacted with combinatorial libraries to generate a multitude of derivative structures, each with different substituent groups. One can see how a vast number of compounds can be generated as a result of the combinatorial process. If a scaffold contains three derivatization sites and the library contains ten groups per site, theoretically 1000 different combinations are possible.

Structure based design, often called rational drug design, is much more focused than combinatorial chemistry. As shown above in Figure above, it involves using the biochemical laws of ligand-receptor association discussed above to postulate ligand refinements to improve binding. For example, we discussed that steric complementarity is vital to tight receptor binding.

Using the crystal structure of the complex, we can target regions of the ligand that fit poorly within the active site and postulate chemical changes to improve complementarity with the receptor. In a similar fashion, functional groups on the ligand can be changed in order to augment electrostatic complementarity with the receptor. However, the danger in altering any portion of the ligand is the effect on the remaining ligand structures. Modifying even a single atom in the middle of the ligand can drastically change the shape of the overall structure. Even though complementarity in one portion of the ligand might be improved by the chemical revision, the overall binding might be severely compromised. This is the difficulty in any ligand refinement procedure.

COMPUTERS IN DRUG DESIGN

There was a great deal of optimism that computer aided drug design would revolutionize the way in which drugs could be developed. The enduring exponential increase in computing power had progressed to the point that rudimentary estimations of ligand receptor complementarity could be performed. Furthermore, computer graphics technology had achieved the ability to generate vector models of chemical structures and manipulate them in real-time. This offered, for the first time, the ability to interactively study computer models of ligand structures and their binding interactions with a receptor.

Concomitant with the development of this technology was the emergence of the AIDS epidemic.Scientists had isolated the causative agent of AIDS, the HIV-1 virus. Considerable characterization of the viral life cycle provided numerous potential targets for pharmaceutical intervention. Among them was the HIV-1 protease. This aspartyl protease was an enzyme that was unique to HIV, and absolutely required for the processing and maturation of HIV proteins. Thus, if a drug could be developed to inactivate this protease, the virus would be unable to generate mature infectious particles to sustain the infection. Numerous groups around the world rapidly solved the crystal structure of this enzyme. The mechanism of this enzyme was determined, and the layout of the active site was carefully mapped.

It was known that humans possess similar classes of proteases. Renin is an enzyme secreted by the kidneys that is responsible for initiating a cascade of reactions that regulate blood pressure. It too is an aspartic protease, and ligands that inhibited its function were known. With the wealth of data from the study of the HIV-1 protease, the hope was that this target could be exploited by computer-aided rational drug design to rapidly generate novel AIDS drugs. Computational chemists believed they could circumvent much of the time and effort required for drug synthesis and testing by simply generating novel compounds using the computer. Testing would be replaced by merely calculating the ligand-receptor binding affinity using the physical laws of chemistry. The concept of generating virtual lead compoundsentirely through computer simulation was termed *denovo design.*

DIFFICULTIES IMPLEMENTING DENOVO DESIGN

Many of the worlds largest pharmaceutical firms spent millions of dollars on hardware and software in their endeavor to make *denovo* design a reality. Unfortunately, successes were rare. Except for a few exceptions, *denovo* design was an utter failure, and did not prove to be an effective method to discover lead compounds. The main reasons were limitations in computing power and the lack of useful software functionality. In scientific computing, accuracy and processing time are always a tradeoff. Thus, in order to make the calculations run in a finite period of time, a plethora of assumptions, significant approximations, and numerous algorithmic shortcuts had to be utilized. This, in turn, greatly diminished the calculated accuracy of any ligand receptor interaction. As such, chemists could postulate numerous chemical structures that could potentially complement the active site; however, the calculated binding had no correlation with reality.

This remains the most significant challenge in *denovo* design to this day. Although computers have become exponentially faster, the sheer number of calculations needed to accurately predict the binding of a *denovo* generated ligand to its receptor in a useful timeframe still requires significant approximations. In *denovo* design, we are attempting to generate a whole ligand from scratch and dock it within the receptor. As stated above, the difficulty lies in predicting how the chemical structure will behave in real life. A ligand is an inherently flexible structure, and can assume a plethora of different conformations and orientations. The big question remains whether the predicted binding structure will mirror the calculated one. Failure in this endeavor has undermined the utility of *denovo* structure generating software. We will discuss these shortcomings and the technological advances of RACHEL, which attempt to circumvent these deficiencies, in detail below.

The second most significant problem in computer aided *denovo* design is the generation of undesired chemical structures. There are a nearly infinite number of potential combinations of atoms. However, the vast majority of these structures are of no use. As discussed above, undesired structures are rejected due to toxicity, chemical instability, or synthetic difficulty. Nearly all *denovo* design software packages are plagued by this problem, especially with respect to synthetic feasibility. Thus, although such software can postulate potential complementary ligands, the vast majority of them are worthless.

The end result of these shortcomings was that computer aided *denovo* design soon fell out of favour as a means of generating viable lead compounds. By then there had been a tremendous number of *denovo* software packages released; however, they all suffered these same problems. Gradually, such programs were shelved and investigators looked to other technologies to aid in their drug development efforts.

Drug discovery and development is an intense, lengthy and an interdisciplinary endeavor. Drug discovery is mostly portrayed as a linear, consecutive process that starts with target and lead discovery, followed by lead optimization and pre-clinical *in vitro* and *in vivo* studies to determine if such compounds satisfy a number of pre-set criteria for initiating clinical development. For the pharmaceutical industry, the number of years to bring a drug from discovery to market is approximately 12-14 years and costing upto \$1.2-\$1.4 billion dollars. Traditionally, drugs were discovered by synthesizing compounds in a time-consuming multi-step processes against a battery of *in vivo* biological screens and further investigating the promising candidates for their pharmacokinetic properties, metabolism and potential toxicity. Such a development process has resulted in high attrition rates with failures attributed to poor pharmacokinetics (39%), lack of efficacy (30%), animal toxicity (11%), adverse effects in humans (10%) and various commercial and miscellaneous factors. Today, the process of drug discovery has been revolutionized with the advent of genomics, proteomics, bioinformatics and efficient technologies like, combinatorial chemistry, high throughput screening (HTS), virtual screening, *de novo* design, *in vitro, in silico*ADMET screening and structure-based drug design.

SILICO DRUG DESIGN

In silico methods can help in identifying drug targets via bioinformatics tools. They can also be used to analyse the target structures for possible binding/active sites, generate candidate molecules, check for their drug likeness, dock these molecules with the target, rank them according to their binding affinites, further optimize the molecules to improve binding characteristics

The use of computers and computational methods permeates all aspects of drug discovery today and forms the core of structure-based drug design. High-performance computing, data management software and Internet are facilitating the access of huge amount of data generated and transforming the massive complex biological data into workable knowledge in modern day drug discovery process. The use of complementary experimental and informatics techniques increases the chance of success in many stages of the discovery process, from the identification of novel targets and elucidation of their functions to the discovery and development of lead compounds with desired properties. Computational tools offer the advantage of delivering new drug candidates more quickly and at a lower cost.

Major roles of computation in drug discovery are:

- Virtual screening and de novo design,
- In silicoADME/T prediction and
- Advanced methods for determining protein-ligand binding.

SIGNIFICATION IN SILICO DRUG DESIGN

As structures of more and more protein targets become available through crystallography, NMR and bioinformatics methods, there is an increasing demand for computational tools that can identify and analyse active sites and suggest potential drug molecules that can bind to these sites specifically. Also to combat life-threatening diseases such as AIDS, Tuberculosis, Malaria etc., a global push is essential. Millions for Viagra and pennies for the diseases of the poor is the current situation of investment in Pharma R&D. Time and cost required for designing a new drug are immense and at an unacceptable level. According to some estimates it costs about $880 million and 14 years of research to develop a new drug before it is introduced in the market Intervention of computers at some plausible steps is imperative to bring down the cost and time required in the drug discovery process

STRUCTURE BASED DRUG DESIGN

The crystal structure of a ligand bound to a protein provides a detailed insight into the interactions made between the protein and the ligand. Structure designed can be used to identify where the ligand can be changed to modulate the physicochemical and ADME properties of the compound,by showing which parts of the compound are important to affinity and which parts can be altered without affecting the binding. The equlibrium between target and ligand is governed by the free energy of the complex compared to the free energy of the individual target and ligand. This includes not only the interaction between target and ligand but also the solvation and entropy of the three different species and the energy of the conformation of the free species.

Virtual screening and de novo design play an important role within the pharmaceutical industry in lead discovery process. Virtual screening refers to computational screening of large libraries of chemicals for compounds that complement targets of known structure which could be tested experimentally. Since, the virtual screening takes place in the three-dimensional active site of the target, it is also called as structure-based virtual screening.

De novo design attemps to use the unliganded structure of the protein to generate novel chemical structure that can bind. There are varying algorithms, most of which depend on identifying initial putative sites of interaction that are grown into complete ligands.

Fragment based discovery is based on the premise that most ligands that bind stongly to a protein active site can be considered as a number of smaller fragments or functionalities. Fragmnents are identified by screening a relatively small library of molecule by X-ray crystallography, NMR spectroscopy.These structues of the fragment binding to the protein can be used to design new ligands by adding functionality to the fragments or by incorporating features of the fragment onto existing ligands.

The phrase "drug-like" generally means molecules which contain functional groups and/or have properties consistent with the majority of known drugs. Lead structures are ligands that typically exhibit suboptimal target binding affinity. Studies have shown that there exists a difference between leads and drugs which can be expressed as follows: Lead structures exhibit, on average, less molecular complexity (less molecular weight, less number of rings and rotatable bonds), are less hydrophobic (lower ClogP and LogD74) and have lower polarizability (less CMR). Leads should display the following properties to be considered for further development in the drug discovery process or to be called as "drug-like":

- Relatively simple chemical features, amenable for combinatorial and medicinal chemistry optimization efforts;
- Membership to a well established SAR (structure-activity relationship) series, wherein compounds with similar structures exhibit similar target binding affinity;
- Favourable patent situation; and
- Good ADME (absorption, distribution, metabolism and excretion) properties.

Leads discovered using virtual screening and *de novo* design methodologies needs to be optimized to produce candidates with improved bioavailability and low toxicity. Studies have indicated that poor pharmacokinetics and toxicity are the most important causes of high attrition-rates in drug development and it has been widely accepted that these areas should be considered as early as possible in the drug discovery process, thus improving the efficiency and cost-effectiveness of the industry. Resolving the pharmacokinetic and toxicological properties of drug candidates remains a key challenge for drug developers.

In silico prediction of drug-likeness at an early stage involves evaluation of various ADMET properties using computational approaches like QSAR or molecular modeling. A number of studies have been performed to find out the properties, which make a drug distinct from other chemicals. Availability of large databases of drug or drug-like molecules, *e.g.* CMC (Comprehensive Medicinal Chemistry), MDDR (MACCS-II Drug Data Report), WDI (World Drug Index) provides useful information about the properties of drugs.

The most influential study of "Lipinski's rule-of-five" identifies several critical properties that should be considered for compounds with oral delivery as concern. A deeper understanding of the relationships between important ADME parameters and molecular structure and properties is needed to develop better *in silico* models to predict ADMET properties. Some of the ADME properties evaluated using *in silico*models are; intestinal permeability, aqueous solubility, human intestinal absorption, human oral bioavailability, active transport, efflux by P-glycoprotein, blood-brain barrier permeation, plasma protein binding, metabolic stability, interactions with cytochrome P450s and toxicity.

NANORESEARCH IN DRUG

There are two major types of drug design. The first is referred to as ligand-based drug design and the second, structure- based drug design.

Ligand Based

Ligand-based drug design (or indirect drug design) relies on knowledge of other molecules that bind to the biological target of interest. These other molecules may be used to derive a pharmacophorewhich defines the minimum necessary structural characteristics a molecule must possess in order to bind to the target. In other words, a model of the biological target may be built based on the knowledge of what binds to it and this model in turn may be used to design new molecular entities that interact with the target.

Structure Based

Structure-based drug design (or direct drug design) relies on knowledge of the three dimensional structure of the biological target obtained through methods such as x-ray crystallography or NMR spectroscopy. If an experimental structure of a target is not available, it may be possible to create a homology model of the target based on the experimental structure of a related protein. Using the structure of the biological target, candidate drugs that are predicted to bind with high affinity and selectivity to the target may be designed using interactive graphics and the intuition of a medicinal chemist. Alternatively various automated computational procedures may be used to suggest new drug candidates.

As experimental methods such as X-ray crystallography and NMR develop, the amount of information concerning 3D structures of biomolecular targets has increased dramatically. In parallel, information about the structural dynamics and electronic properties about ligands has also increased. This has encouraged the rapid development of the structure-based drug design. Current methods for structure-based drug design can be divided roughly into two categories. The first category is about "finding" ligands for a given receptor, which is usually referred as database searching. In this case, a large number of potential ligand molecules are screened to find those fitting the binding pocket of the receptor.

This method is usually referred as ligand-based drug design. The key advantage of database searching is that it saves synthetic effort to obtain new lead compounds. Another category of structure-based drug design methods is about "building" ligands, which is usually referred as receptor-based drug design. In this case, ligand molecules are built up within the constraints of the binding pocket by assembling small pieces in a stepwise manner. These pieces can be either atoms or fragments. The key advantage of such a method is that novel structures, not contained in any database, can be suggested. These techniques are raising much excitement to the drug design community.

ACTIVE SITE IDENTIFICATION

Active site identification is the first step in this program. It analyses the protein to find the binding pocket, derives key interaction sites within the binding pocket, and then prepares the necessary data for Ligand fragment link. The basic inputs for this step are the 3D structure of the protein and a pre-docked ligand in PDB format, as well as their atomic properties. Both ligand and protein atoms need to be classified and their atomic properties should be defined, basically, into four atomic types:

- *Hydrophobic atom*: All carbons in hydrocarbon chains or in aromatic groups.
- *H-bond donor*: Oxygen and nitrogen atoms bonded to hydrogen atom(s).
- *H-bond acceptor*: Oxygen and sp2 or sp hybridized nitrogen atoms with lone electron pair(s).
- *Polar atom*: Oxygen and nitrogen atoms that are neither H-bond donor nor H-bond acceptor, sulfur, phosphorus, halogen, metal and carbon atoms bonded to hetero-atom(s).

The space inside the ligand binding region would be studied with virtual probe atoms of the four types above so the chemical environment of all spots in the ligand binding region can be known. Hence we are clear what kind of chemical fragments can be put into their corresponding spots in the ligand binding region of the receptor.

SCORING METHOD

Structure-based drug design attempts to use the structure of proteins as a basis for designing new ligands by applying accepted principles of molecular recognition. The basic assumption underlying structure-based drug design is that a good ligand molecule should bind tightly to its target. Each component reflects a certain kind of free energy alteration during the binding process between a ligand and its target receptor. The Master Equation is the linear combination of these components. According to Gibbs free energy equation, the relation between dissociation equilibrium constant, Kd and the components of free energy alternation was built.

The sub models of empirical functions differ due to the consideration of researchers. It has long been a scientific challenge to design the sub models. Depending on the modification of them, the empirical scoring function is improved and continuously consummated

COMPUTER-AIDED DRUG REFINEMENT

It was at this time that the techniques of mass screening and combinatorial chemistry began to gain widespread acceptance and use.

The use of mass screening and combinatorial chemistry allowed researchers to discover lead compounds in a rapid and efficient manner. As such, *denovo* design tools and their associated problems were no longer needed to generate lead structures. One would surmise that computer-aided drug design technology would have soon ceased to exist. On the contrary, it soon became apparent that computational tools were needed that could optimize these lead compounds into potent drugs.

The concept of drug optimization versus *denovo* design is an important one. The difficulty with *denovo* ligand generation is that an entire structure is being created from scratch. The confidence one has of accurately predicting how this structure will interact and bind within a target receptor is shaky at best. In drug optimization, we begin with a lead compound whose bound structure within the receptor has been characterized, most likely through x-ray crystallography.

Subtle modifications are then performed to generate derivative compounds using structure based drug design to improve binding affinity. Because we are making much smaller changes, our faith in the validity of the resulting structures is far greater. These derivatives then undergo testing to determine which modifications improve binding. The structures of the best ligands can then be elucidated to verify the accuracy of the modifications. This refinement process continues iteratively until optimal binding ligands are produced.

Since subtle modifications are being made to a common structure, the predictive ability of ligand refinement software is much higher. This is because the effect of a single chemical modification on ligand-receptor binding is far easier to quantitate than an extreme change. No longer are we trying to determine the binding affinities of drastically different structures. Instead, we are simply determining the rank order of a list of derivative compounds. This greatly increases the confidence that proposed structures will bind in a manner consistent with our understanding.

In addition, the act of generating chemical derivatives is highly amenable to computerized automation. Consider the application of targeted structure based combinatorial chemistry as discussed above. Libraries of derivative components are assembled based upon the analysis of the active site. Because of the combinatorial nature of this method, an extremely large number of candidate structures may be possible.

A computer can rapidly generate and predict the binding of all potential derivatives, creating a list of the best potential candidates. In essence, the computer filters all weak-binding compounds, allowing the chemist to focus, synthesize, and test only the most promising ligands. Thus, utilizing computer aided drug design software to aid in the refinement of weak binding lead compounds is the most effective manner in which these tools can be employed. The use of computer modeling to refine structures has become standard practice in modern drug design.

PRINCIPLES DRUG FORMULATION

The formulation of a drug compound, *i.e.* the way in which is it combined with excipients (such as: binders, diluents, solubilizers and disintegrants) in, for example, tablets, capsules, creams pastes and gels can be critical to the performance of the active ingredient. The formulation must also be fully characterized to ensure product efficacy and patient safety. Techniques used for characterization include X-ray crystallography, vibrational spectroscopy, thermal analysis, solid state NMR and microscopy.

Polarized light microscopy, in particular, can be used to study a wide range of physicochemical properties. Materials can be characterized and identified by optical properties such as refractive index and birefringence and other phenomena such as particle size, particle shape, colour, twinning and the presence of inclusions or contaminants. When using a heated stage further temperature related properties of materials may be investigated, such as polymorphic transitions (for example, crystallization and melting points).

During storage, it is essential that formulations are not susceptible to microbial contamination. Light microscopy can be used to determine the effectiveness of preservatives and antimicrobials. Many contaminants can be characterized using appropriate staining techniques.

METABOLISM

The term ADME is typically used in reference to nonclinical studies. In reality, ADME is generic, being applicable without restriction to pharmacokinetic/metabolism investigations in humans as well as animals. Fundamentally, ADME information is critical in all phases of a fully integrated drug development program. The initial charge of early development programs is to file the Investigational New Drug (IND) application such that approval may be secured to investigate a new chemical entity (NCE) in humans.

Animal toxicology studies comprise the foundation of the IND. The toxicology program is designed to identify toxicities that may potentially be encountered in humans, helping ensure that the initial human studies are conducted safely and ethically. In this context, ADME studies provide supportive information to augment the interpretation of toxicology findings. Of primary importance among them are toxicokinetic studies, in which systemic drug and/or metabolite exposure in toxicology animals is evaluated. Drug exposure, expressed in terms of AUC (area under the drug plasma concentration-time curve), Cmax (maximum drug concentration in plasma), or an alternative parameter, is then related to dose level and toxicological outcomes. Based on toxicokinetic data at the no-observed toxic effect dose, an acceptable exposure limit in humans can be defined. ToxicokineticsPharmacokinetics/absolute bioavailability in toxicology species (male/female)Protein bindingErythrocyte/plasma distributionWhole body autoradiography/tissue distributionMass balance in toxicology speciesMetabolite profile in toxicology speciesPharmacodynamicsAllometric scaling

In later development, the focus of ADME work is shifted to human studies. These are intended to more fully define the disposition of the drug in humans, particularly in the target therapeutic population. Ultimately these data are integrated into a New Drug Application (NDA) or Market Authorization Application

ADME IN THE DRUG DISCOVERY PROCESS

While scientists in the ADME disciplines have historically provided support to drug discovery teams, the unique expertise, which they could bring to the discovery process, has heretofore been vastly underutilized. In no small way this may be ascribed to the less than expedient experimental procedures required to conduct most ADME studies. The time- and labour-intensiveness of much of this work is not an issue during drug development, where resources can be focused on one compound. This *modus operandi* does not conform, however, to current drug discovery efforts, which have exponentially enhanced their operating efficiency with mass-screening techniques. However, the advent of a host of exciting new technologies is providing hope that scientists in various of the ADME disciplines may now be able to participate more actively and productively in the drug discovery process.

Absorption

Insofar as the overwhelming majority of new drugs are

intended to be administered orally, with few exceptions the ability of an NCE to provide activity by the oral route (*i.e.* have good bioavailability) is imperative. Whereas oral bioavailability can be determined relatively easily in animal pharmacokinetic studies, it is not reasonable to expect that this approach can function optimally in a large scale-screening environment. Fortunately a number of *in vitro* and cell culture techniques have evolved in recent years that have facilitated the assessment of intestinal permeability.

It has been demonstrated that membrane permeability can be predicted for some compounds with reasonable accuracy based solely on physicochemical parameters. Therefore, close scrutiny of the chemical structure may provide valuable basic information about an NCE prior to the commencement of laboratory experiments. It is well established, for instance, that efficient oral absorption will occur only after drug has dissolved and presented itself to the intestinal mucosal surface from whence it can traverse the epithelium. Dissolution is determined by the highly interdependent influences of aqueous solubility, ionizability (pKa), and lipophilicity (octanol/water log P or log D7.4). Furthermore, log P is a crucial factor governing passive membrane partitioning, influencing permeability opposite to its effect on solubility (*i.e.* increasing log P enhances permeability while reducing solubility). In light of this counterdependence, it has been suggested that oral absorption may be optimal within a log P range of 0.5 to 2.0.

While the promise of predicting membrane permeability from chemical structure alone is enticing, these methods do not yet enjoy the level of sophistication required to supplant experimental methods. In that regard, a number of *in vitro* tools have been adapted, with notable efficiency, to high-throughput assessments of membrane permeability and potential oral bioavailability. Most notable among them are CaCO-2 cells. Derived from a human colon carcinoma cell line, these cells are grown in a confluent monolayer on porous membrane filters which are mounted in diffusion chambers. Permeability measurements are based on the rate of appearance of test compound in the receiver compartment. The apical (donor) surface of the monolayer contains microvilli and thus retains many characteristics of the intestinal brush border. The cells also express functional transport proteins and metabolic enzymes, the degree of expression being dependent upon the post-seeding age of the cells.

Everted intestinal rings and brush-border membrane vesicles (BBMV) are also commonly used systems for assessing membrane permeability. The former technique is a refinement of one of the earliest *in vitro*absorption systems in which an everted intestinal segment was suspended in a buffer system to measure mucosal-to-serosal transfer. The current methodology involves isolating a rat intestinal segment, everting, and slicing into rings, which are suspended in buffer. Removing the brush-border surface from rat or rabbit intestine and molding it into vesicles by homogenization and differential centrifugation prepare BBMV. Both everted intestinal rings and BBMV are most useful for determining drug uptake rates rather than transepithelial flux.

In the *in situ* intestinal perfusion system, an intestinal segment is exposed in an anesthetized rat and drug solution is perfused through the lumen in a single-pass or recirculating fashion. Drug permeability is derived from the rate of disappearance of drug from the perfusate. Though more labour-intensive, in situ intestinal perfusions remain popular owing to the perceived clinical relevance of permeability data derived therefrom. In a recent study, it was demonstrated with a series of small organic molecules as well as a series of peptidomimetics that CaCO-2 cells, everted intestinal rings, and in situ perfusions have strong potential for predicting fraction absorbed in humans.

One of the most appealing attributes, which these experimental systems possess, is their capability to perform relatively high throughput screening. This is particularly relevant for CaCO-2 cells, everted intestinal rings, and BBMV. An imposing rate-limiting step, once the systems are established and optimized, is the development of assay methods (usually HPLC) to quantify the analytes of interest. On the positive side, as these experiments are conducted in aqueous buffers, the preanalytical sample purification requirement is minimal. These analytical burdens can be even further reduced by immobilized artificial membranes (IAM), a recently developed system which has been greeted with considerable enthusiasm .

Briefly, the analyte is injected onto a specialized IAM HPLC column packed with a phosphatidylcholine (PC) stationary phase selected to closely mimic biological membranes. Theoretically, the chromatographic capacity factor for the analyte should correlate with the water-to-PC membrane partition coefficient, a useful parameter for calculating membrane permeability. The ability of IAM technology to predict membrane permeability is currently being evaluated. Elegant in its simplicity, it is well suited to a rapid screening environment, although it is recognized that its applicability will be limited to those compounds, which are absorbed by passive processes.

Metabolism

The systems described above have the greatest utility when absorption is rate limiting to systemic availability. For many compounds, even if absorption is optimized bioavailability may be limited by extensive metabolism. Indeed, metabolism can complicate the *in vivo* activity profile irrespective of route of administration.

As with absorption, valuable metabolic input can be imparted to the drug discovery team based on information gleaned from chemical structure alone. It is felt that the teleological underpinnings of xenobiotic transformation reside in modifying environmental toxins so as to facilitate their removal from the body. Not surprisingly, the majority of drug biotransformation processes appear to follow this paradigm, typically resulting in increasingly water-soluble metabolic products which can be more readily excreted in urine. Thus one may surmise that within a group of compounds, highly lipophilic drugs would be metabolized most actively. More explicitly, the metabolic potential of a compound containing a chemical moiety known to be avidly biotransformed may be predicted reasonably accurately on the basis of abundant historical data. For example, being cognizant that unhindered phenolic hydroxyl groups are exquisitely good substrates for conjugating enzymes, NCEs, which contain this moiety, can be eliminated from further consideration if a predilection to rapid metabolism is felt to be detrimental. As we gain information about the specific conformational constraints imposed by the catalytic sites of major metabolic enzymes, our ability to predict *in vivo* metabolic events based solely on chemical structure is enhanced commensurately.

Given the current state of knowledge, however, it is likely that drug discovery teams will continue to prefer metabolism-related input predicated on experimental data rather than theoretical predictions. To this end, various *in vitro* methods are available which are being increasingly incorporated into drug discovery strategies. Among the most popular and widely utilized systems in use today are hepatic microsomes. These preparations retain activity of those enzymes, which reside in the smooth endoplasmic reticulum, such ascytochromes (CYP), flavin monooxygenases, and glucuronosyltransferases.

Isolated hepatocytes appear to retain a broader spectrum of enzymatic activities, including not only reticular systems, but cytosolic and mitochondrial enzymes as well. Because of a rapid loss of hepatocyte-specific functions, it has been possible to generate useful data only with short term hepatocyte incubations or cultures. However, significant strides have been made towards maximizing the viability of hepatocytes in culture.

Liver slices, which like hepatocytes retain a wide array of enzyme activities, are also increasing in popularity. Furthermore, both hepatocytes and liver slices are capable of assessing of enzyme induction *in vitro*. The choice of which system to employ in a drug discovery screening program will depend on many factors, not the least of which is availability of tissue for large-scale implementation. In addition, historical information about a particular chemical series is invaluable. Consider, for example, an SAR-driven chemical series with a prototype analogue that has been demonstrated to induce CYP activity. For metabolic screening in this program, hepatocytes or liver slices may be preferable to microsomes. Biotransformation of NCE candidates as well as induction potential can be evaluated in the same system, making the added resources required to establish these systems well worth the investment. Information gathered from *in vitro* metabolism studies is especially useful in choosing drug candidates for future development. The potential utility of *in vitro*metabolism data in predicting *in vivo* intrinsic clearance has been touted.

Isolated heterologous human CYP enzymes have been available for several years, being expressed from cDNA in yeast (*Saccharomyces* cerevisiae), bacterial (*Escherichia coli*), and mammalian (B-lymphoblastoid) cell lines. These systems can be used to ascertain whether a compound is a substrate for a particular CYP isozyme and, if so, what metabolite is generated by that enzyme.

Moreover, these enzymes, in sufficient quantity, may possibly be used as bioreactors to generate usable amounts of a metabolic product that may be difficult to chemically synthesize in the laboratory. Isozyme-specific antibodies and isozyme-specific inhibitory substrates, by selectively abolishing the activity of a particular CYP isoform, may be used to determine the relative importance of that enzyme in the turnover of an NCE. Conversely, the ability of an NCE to interact with a particular CYP isozyme may be determined by metabolic cross-inhibition studies against prototypical CYP substrates.

In vitro metabolism systems are not limited to those derived from the liver. Most pharmaceutical companies have been building liver and tissue banks to permit a cross-comparison of metabolic turnover rates in various tissues from various species. It is not uncommon now for liver banks to house tissues from a variety of species, including those from animals treated with enzyme inducers or inhibitors. Therefore, cross-species *in vitro* metabolism comparisons are becoming more feasible and commonplace. In addition to providing information on potential rates and routes of metabolism, interspecies comparisons may help in choosing species to be used in toxicology studies.

Available *in vitro* metabolism technologies are considerably more efficient than traditional *in vivo* methods. With these systems, it is possible to assess the relative rates and routes of biotransformation of a handful of compounds in the time required to comparably characterize one compound using *in vivo* approaches. In the past, a relative paucity of tissues was the most notable impediment to conducting large-scale screening efforts with *in vitro* tools. However, the continued growth of in-house tissue banks, in concert with the optimization of isolated enzyme expression systems, has made high-throughput metabolism assessments in the discovery arena a reality.

PHARMACOKINETICS

Unfortunately, there is no substitute for actual *in vivo* data in assessing pharmacokinetic profiles of drug candidates. While insight into various aspects of the pharmacokinetic profile (absorption, metabolism, protein binding) can be gleaned from *in vitro* techniques, there are as yet no methods available for accurately predicting what will happen to a drug when it is put into a whole animal.

For a useful assessment of pharmacokinetics and bioavailability, it is necessary to administer the drug to selected animal species both intravenously and by the intended route of administration (usually oral). Blood samples are collected over a predetermined time course after dosing and the drug is quantified in serum or plasma by a suitable bioanalytical method (*e.g.* HPLC). Alternatively or concurrently it may be possible to collect plasma samples from animals used in whole animal pharmacologic models and, based on the concentration/effect (pharmacokinetic/pharmacodynamic) relationship established, make a link between *in vitro* pharmacologic activity and the behaviour of a compound *in vivo*.

In any case, the most significant impediment to providing pharmacokinetic input to a high-throughput discovery team is the time and labour-intensiveness of the bioanalytical methods available. To assess biochemical or pharmacologic activity *in vitro* or *in vivo*, a standardized screening method is established with a common assay endpoint that can be applied to test all compounds that are available. In contrast, a separate bioanalytical method for pharmacokinetic assessment must be developed for each compound.

For chromatographic assays, which comprise the vast majority of the methods employed in pharmacokinetic studies, remarkable improvements in assay detection limits and sample cleanup have been realised. However, little has been done to shorten the time required for assay development and implementation. Pharmacokinetic characterization is therefore often relegated to end-stage discovery, being utilized to select which of 2 or 3 potential lead candidates has the most "acceptable" pharmacokinetic profile. Clearly, if pharmacokinetic input is to be available for early discovery decisions, more efficient methodologies are necessary.

Semi-simultaneous bioavailability estimation is a screening method which has been successfully utilized in drug discovery. With this technique, an intravenous dose of drug is administered and at a suitable time postdose (*i.e.* postdistributional) an oral dose is administered to the same animal. Pharmacokinetic parameters such as clearance and volume of distribution can be determined from the intravenous concentration-time curve. Oral pharmacokinetic parameters, including bioavilability, are subsequently extricated from the combined IV and PO concentration-time data by deconvolution. By reducing inter- and intraanimal variability, accurate pharmacokinetic data can be gathered with fewer animals. In addition, the total number of plasma samples from both routes of administration is reduced by the overlapping dosing regimes. On the other hand, a prior knowledge of the pharmacokinetic characteristics of the NCE is needed to optimize the administration paradigm. In that regard, it may be possible to extrapolate from studies with structurally analogous predecessors from the same chemical series. More critical, however, is that compound-specific analytical methods are still required for each NCE.

Bioassays represent one alternative to compound-specific assays. Plasma samples are added to an *in vitro*system and the biological activity is quantified in much the same way as a pure drug solution is tested for activity. The extent of displacement of a prototype receptor ligand or inhibition of a target enzyme activity is measured as the assay endpoint for each sample. Based on standard curves, which relate the assay endpoint to known quantities of added drug, the concentration of the compound in the plasma can be quantified. Microbiological assays, which measure the inhibition of bacterial growth on an agar plate in relation to the amount of antibacterial drug applied, are among the most widely utilized examples of this technique.

The greatest advantage of bioassays over HPLC methods is that they can be employed for analysing any compound which possesses the appropriate biological activity, making them amenable to high throughput screening. Unfortunately, owing to their non-specific nature, these assays will quantify all biological activity in a sample, which could include active metabolites. These results may be misleading if one is attempting to understand the pharmacokinetic behaviour of only parent compound. In response, it could be argued that it is prudent to choose the compound, which provides the best activity profile, irrespective of whether the in vivo activity observed, is due to parent compound or active metabolite. Preprocessing of plasma samples (extraction) could possibly remove polar metabolites, allowing the bioassay to concentrate on parent drug.

In perhaps the most significant technologic leap in this field, scientists are capitalizing on the exquisite selectivity of mass spectrometry (MS) to circumvent the tedium of HPLC assay development. The resulting procedure is the pharmacokinetic equivalent of a mass-screen, the simplicity and elegance of which is the result of several evolutionary steps. In its earliest incarnation, animals were dosed with a single drug and samples were collected, processed, and injected onto an HPLC- MS. The chromatographic system was optimized to isolate the drug peak and the mass spectrometer was used to detect the appropriate mass ion and quantify it. Eventually, HPLC separation was foregone in favour of direct injection of a minimally preprocessed plasma sample into the MS. In a further refinement, multiple drugs were boldly administered simultaneously to the same animal and each compound was then individually quantified.

MS is capable of distinguishing subtle differences in the characteristic mass ion peaks of each analyte, imparting a degree of specificity that HPLC cannot achieve without fairly elaborate intervention. Under this protocol, the number of drugs for which pharmacokinetic profiles can be obtained in a week increases several-fold, being limited only by the number of compounds, which are co-administered in each animal. The considerable sample burden thus engendered is easily accommodated by the facile, direct-injection MS analysis. Overcoming myriad theoretical reasons why it shouldn't work (*e.g.* potential for metabolic interactions, protein binding displacement, or additive pharmacologic effects), investigators in several large pharmaceutical companies have successfully incorporated this screening approach into routine drug discovery support.

Drug distribution and tissue penetration data are occasionally obtained during discovery screening. Plasma protein binding assessments can be relatively easily performed using well-known and widely-used methods, providing perspective on the relationship between total *in vivo* plasma drug concentrations and those of pharmacologically available unbound drug.

For more specific applications, cell cultures techniques are available for predicting the ability of compounds to traverse target biological membranes (*e.g.* blood-brain barrier, T lymphocytes). *In vivo* microdialysis has been used for assessing drug penetration into CNS and liver, but is methodologically difficult to transform into a screening tool. In addition, these methodologies typically suffer from the same bioanalytical limitations of classical pharmacokinetic profiling. It seems reasonable to hope that direct-injection MS analysis may facilitate these types of studies in the future.

ADME IN DRUG DESIGN

With continued refinement and streamlining of its underlying technologies, ADME-relateddiscovery screening can be implemented ever earlier into the discovery process. Ultimately, this will permit scientists in the ADME disciplines to play a significant role inoptimizing new drug design. Fundamentally, that role is to identify the limiting barriers to optimal *in vivo* efficacy.

Of all the attributes, which determine a drug's ultimate *in vivo* efficacy, physicochemical behaviour is perhaps the most basic. For this reason alone it deserves considerable scrutiny by the drug discovery team. Moreover, as solubility, pKa, and lipophilicity are so integrally linked with chemical structure, physicochemical properties in some respects can be relatively predictably manipulated by chemical modifications. Historically, chemical synthetic efforts have been guided almost exclusively by SAR intended to maximize interaction with a therapeutic target (*e.g.* receptor, enzyme). The goal of ADME-guided synthesis, on the other hand, is to maximize the ability of an NCE to access the therapeutic target *in vivo*. As these synthetic directives may often be at odds, the Holy Grail for discovery teams in the future will be to achieve an optimal balance between the interests of these groups and fully integrate the valuable information which they can each impart towards guiding new drug design.

Lipophilicity is a critical determinant of a host of ADME processes. It has been suggested, for example, that compounds with log P values between 0 and 3 or 4 (*e.g.* 0.5 to 2) are the most suitable candidates for passive transcellular absorption across intestinal epithelia. As lipophilicity is increased (log P greater than 3 or 4), solubility and hence absorption progressively decline. In contrast, more hydrophilic compounds with log P values less than 0 are likely to traverse the epithelium more slowly via paracellular channels, although it should be appreciated that molecular size is also an important determinant of transit through the narrow (5-9 Å) paracellular channels. Similarly, it has been observed that compounds with log P values greater than 0 are likely to undergo substantial renal tubular reabsorption. By itself, this effect will tend to prolong $t_{1/2}$, however the greater lipophilicity and minimal renal excretion will also render the compound more susceptible to metabolism.

Experimental determination of lipophilicity is not a trivial matter, a fact which has limited the availability of this information in early discovery where it could be most useful. However, mathematical predictions based on a weighting of functional groups within a molecule have been performed with impressive accuracy. As confidence in the reliability of these predictions is further developed, it will become more feasible in the synthesis of new drugs to target log P values to achieve a balance between particular ADME processes.

Even more so than log P, pKa is a physicochemical property that can be readily predicted, given the vast experience amassed over many years with myriad ionizable organic functional groups. The ionization state of an NCE *in vivo* can be predicted by considering its pKa relative to pH in various body compartments. It is possible to surmise, for example, that an organic acid will present to the nephron in an anionic state, rendering it potentially susceptible to active tubular secretion by the organic anion transporter. If the compound is indeed a substrate for this transporter, renal clearance may be high and elimination from the systemic circulation rapid.

Both log P and pKa are the primary determinants of compound solubility. *In vivo,* this has a decisive bearing on oral absorption, following the premise that drugs must dissolve in the GI tract to be available to cross the intestinal membrane. Given that pH progressively increases throughout the length of the gastrointestinal tract, solubility and thus absorbability of an ionizable compound will vary accordingly. Solubility, log P, and pKa are closely intertwined determinants of renal tubular reabsorption as well. Thus, whereas compounds with log P values greater than 0 may be potential candidates for tubular reabsorption, this process will only be likely to occur if the compound is present in urine (pH 4.5 to 8) in an un-ionized state and has not exceeded its solubility limit. Despite the importance of physicochemical properties in determining the ADME profile, they are as yet primarily rough guidelines and cannot substitute for experimentation to bear out the validity of their predicted *in vivo* effects. Experimental data generated by the discovery tools outlined above can help put the importance of physicochemical properties into the appropriate context.

Structural modifications may also be guided by more fundamental knowledge of a particular ADME process. The stability of peptidomimetic compounds to peptidases in the GI tract may be improved by various synthetic alterations, including protection of peptide bonds with steric bulk or substitution of naturally occurring L-amino acid residues with their D-isomers. The increased GI stability potentially increases bioavailability, but must be balanced against the contribution of physicochemical properties such as lipophilicity. In addition, it is important to note that although lipophilicity and GI stability may favour intestinal absorption, high molecular weight (>600 Da) of some of these compounds could render them acutely sensitive to biliary excretion, reducing absolute systemic bioavailability in spite of their good intestinal permeability.

Among the many transport proteins, which reside in the GI mucosa, the dipeptide transporter has been shown to be an important mediator of active drug absorption. In addition to di- and tripeptides, compounds from several compound classes, which are structurally similar to di- and tripeptides, have also been shown to be substrates for this carrier, including -lactams and ACE inhibitors. Targeting synthesis to enhance the affinity of NCEs for this transporter may prove to be a fruitful approach to improving systemic bioavailability, particularly for polar compounds with inherently low membrane permeabilities. In one such example, formation of a proline adduct improved the transcellular passage of L-methyldopa by transforming it into a substrate for the dipeptide transporter. In addition to being dipeptide-like, L-methyldopa-pro is also a substrate for intracellular prolidase which was specifically exploited to regenerate L-methyldopa after the adduct was transported across the apical surface of the enterocyte. As their substrate specificities are further delineated, other known transporters could be similarly targeted for enhancing bioavailability.

Prodrug strategies have successfully improved the oral bioavailability of numerous compounds. In many cases, this involves masking a polar group by esterification to increase lipophilicity and enhance the extent of absorption from the GI tract. After absorption, the ester is enzymatically hydrolyzed to release parent drug. Absorption has also been improved with prodrugs designed to enhance solubility by decreasing lipophilicity, targeting progroup cleavage for the GI tract. With this approach, the nature of the progroup predetermines whether parent drug is released in the GI lumen or at the brush border membrane.

The ideal candidate for the latter manipulation is a highly lipophilic compound, which readily traverses the GI mucosa, but is too insoluble to produce a reasonable soluble fraction for acceptable bioavailability. A polar progroup (*e.g.* phosphate, amino acid) is incorporated into the molecule, markedly increasing its solubility in the GI tract. The progroup is subsequently cleaved at the intestinal brush border by the targeted enzymes (*e.g.* phosphatases, aminopeptidases), which expels the lipophilic parent compound in close proximity to the intestinal surface. The low local concentration minimizes the possibility of precipitation and allows the drug to be directly absorbed. The bioavailability of several compounds has been improved with this strategy.

Soft drugs provide an interesting counterpoint to prodrugs. In both cases, the administered compound is designed to undergo a predictable metabolic modification. With prodrugs, this biotransformation event rapidly converts an inactive precursor into an active drug. Conversely, a soft drug is itself the active drug, which, after eliciting a therapeutic effect, is metabolically inactivated in a predictable manner and at a controlled rate. Several approaches to rational soft drug design have been identified, including soft analogs, activated soft compounds, natural soft drugs, the active metabolite approach, and the inactive metabolite approach. In the latter case, an inactive metabolite to a known drug is first identified. This metabolite is then chemically modified to create a new drug which:

Possesses comparable pharmacologic activity to the original drug:

- Will be biotransformed to the inactive metabolite in one step and without going through toxic intermediates, and
- May possibly be subjected to further modification if transport or binding properties require optimization.

The metabolic systems typically harnessed by soft drugs are hydrolytic enzymes (*e.g.* esterases). Monoxygenases are scrupulously avoided in the belief that hydrolyses are inestimably more predictable and controllable processes. In reality, it may be difficult to integrate the soft-drug concept into novel, innovative drug discovery programs. Its retrometabolic philosophy is based on the premise that the metabolic fate of a pharmacologically active drug is known in explicit detail. Therefore, unless the synthetic strategy is predicated on a well-established chemical series, the information required for soft-drug design will not be available. However, the basic principles of this concept might possibly be extrapolated from successful soft drugs to novel compounds with similar functional moieties.

The guidance of drug design based on more classical metabolism theory is multifaceted. As with any strategy, an ability to predict the metabolic fate of an NCE based simply on its structure would be of untold value. To that end, several computerized " Expert systems" have been developed which are designed to promulgate for any given compound a list of likely metabolic products. Unfortunately, these systems have proven to be perhaps too general in their scope, projecting an array of potential products so extensive as to be of relatively limited use. On a slightly different tack, a vast metabolite database is currently being developed which is based on metabolism studies reported in the literature over the past 90 years. This package, from Molecular Design, interfaces with

ISIS databases and is thought to hold considerable promise for providing more realistic projections of metabolic routes across species.

Molecular modeling may come to fruition in the near future. Computer assisted drug design might allow the determination of NCE-receptor compatibility to proceed concurrently with an evaluation of the ability of the NCE to fit into the active site of a particular metabolic enzyme. This capacity does not presently exist on any significant scale. It is particularly frustrating with respect to the CYP family, which is the most important of Phase 1 metabolic enzyme systems. In part because CYP is a membrane-bound enzyme in close association with a reductase, it has not yet been crystallized. Therefore, an X-ray structure, required for building a useful computer model, is not available. The multiplicity of the CYP family makes the task even more formidable.

Therefore, guiding synthesis towards an optimal metabolic profile is based as yet on largely empirical insight. Substantial inroads have been made in this regard, particularly with respect to 4 of the most important CYP isozymes, CYP2D6, 2C9, 2C19, and 3A4. Models of the active sites of CYP2D6 and CYP2C19 have been hypothesized based on a comparison of chemical structures of a variety of substrates, taking into consideration the strict stereo- and regioselectivity of their transformations.

The high degree of selectivity inherent in their active sites make it feasible to design drugs which may be more or less likely to interact with these isozymes. In one example, addition of a bulky para-substituent to the basic beta-blocker nucleus ultimately yielded betaxolol, which unlike its pharmacologic brethren is resistant to the action of CYP2D6. CYP3A4, on the other hand, has to date defied attempts to define its active site, being far less discriminating in its choice of substrates. Indeed, some rather large molecules, such as cyclosporin and macrolide antibiotics are known to be turned over by CYP3A4. As metabolism databases expand in the coming years, a better delineation of how to chemically modify NCEs will be forthcoming, permitting more predictable interactions with

CYP isozymes. Some general rules are in place with regard to avoiding chemically reactive structures. Thus, compounds containing structural elements with a high risk of producing toxic products should be strictly avoided. Examples include quinones and quinonimines, aromatic methylene-dioxy groups, and aromatic nitro groups.

Enzyme induction is more difficult to predict. There is little guidance with regard to what chemical structures are known to produce this effect, with the possible exception of broad chemical groups such as barbiturates, hydantoins, di-t-butyl phenols, etc. However, even within such series, subtle changes in chemical structure can markedly and unpredictably alter enzyme induction potential .

As detailed above, pharmacokinetic characterization is often relegated to end- stage discovery, at which point compound attrition has whittled the pool of potential lead compounds to a small, manageable number. Under these circumstances, the opportunity to provide meaningful feedback, which might assist in pharmacokinetically optimizing drug design, is limited. As analytical techniques become more expeditious, pharmacokineticists will play an increasingly important role in drug design. Nonetheless, it should be clearly recognized that pharmacokinetics in and of itself is not particularly meaningful.

Pharmacokinetic behaviour is highly interdependent with absorption and metabolism. However, characterizing the pharmacokinetic profile of a compound is still of utmost importance insofar as it helps put absorption, metabolism, and other data into the proper perspective. For example, *in vivo* activity may be poor despite predictions from *in vitro*absorption and metabolism studies that an NCE will be well absorbed and not be subjected to appreciable biotransformation. The pharmacokinetic profile reveals that the drug has a short *in vivo* half-life owing to a small volume of distribution and high renal clearance, neither of which could have been predicted accurately prior to the whole animal work. In this case, an optimization of the PK profile may be guided by a realization that chemical modifications are needed which will specifically reduce renal clearance or alter plasma protein binding. The ability to predict *a priori* the extent of plasma protein binding is limited, although fruitful attempts have been made to define the molecular binding requirements for certain sites on human serum albumin. In general, binding is likely to increase with lipophilicity; an anionic group may enhance binding to albumin even further. In contrast, cationic molecules are virtually precluded from binding to albumin, rather favouring -1-acid glycoprotein. Ultimately, despite formidable theoretical arguments to the contrary, the role of protein binding in determining *in vivo* efficacy is still not universally predictable. For example, it has been suggested that NCEs with low protein binding are the most ideal candidates for further development. Quite often this implies low lipophilicity, which could be at odds with strategies for optimizing membrane permeability or receptor or enzyme affinity. It has also been suggested that high protein binding is not necessarily a negative attribute and in fact may help prolong the activity of some compounds with otherwise short half-lives.

TOXICITY

Toxic side effects of local anaesthetic drugs occur when excessive blood levels occur.

This is usually due to:

- Accidental rapid intravenous injection.
- Rapid absorption, such as from a very vascular site ie mucous membranes. Intercostal nerve blocks will give a higher blood level than subcutaneous infiltration, whereas plexus blocks are associated with the slowest rates of absorption and therefore give the lowest blood levels.
- Absolute overdose if the dose used is excessive.

REDUCING THE RISK OF TOXICITY

- Decide on the concentration of the local anaesthetic that is required for the block to be performed. Calculate the total volume of drug that is allowed according to the table below.

- Use the least toxic drug available.
- Use lower doses in frail patients or at the extremes of ages.
- Always inject the drug slowly (slower than 10ml/ minute) and aspirate regularly looking for blood to indicate an accidental intravenous injection.
- Injection of a test dose of 2-3ml of local anaesthetic containing adrenaline will often (but not always) cause a significant tachycardia if accidental intravenous injection occurs.
- Most nerve blocks are more dependent on volume of drug injected than the total dose. Therefore if more volume is needed it is better dilute the local anaesthetic with 0.9% saline than to add more local anaesthetic and increase the dose unnecessarily.
- Add adrenaline (epinephrine) to reduce the speed of absorption. The addition of adrenaline will reduce the maximum blood concentration by about 50%. Usually adrenaline is added in a concentration of 1:200,000, with a maximum dose of 200 micrograms.

SIGNS AND SYMPTOMS

The systemic toxic effects due to local anaesthetic overdose primarily involve the central nervous and cardiovascular systems. In general the Central Nervous System (CNS) is more sensitive to local anaesthetics than the Cardiovascular System (CVS). Therefore CNS manifestations tend to occur earlier. Brain excitatory effects occur before the depressant effects.

CVS SIGNS AND SYMPTOMS

- *Early or mild toxicity*: tachycardia and rise in blood pressure. This will usually only occur if there is adrenaline in the local anaesthetic. If no adrenaline is added then bradycardia with hypotension will occur.
- *Severe toxicity:* Usually about 4 – 7 times the convulsant dose needs to be injected before cardiovascular collapse occurs. Collapse is due to the

depressant effect of the local anaesthetic acting directly on the myocardium. Bupivacaine is considered to be more cardiotoxic than lignocaine. Severe and intractable arrhythmias can occur with accidental iv injection.

The acute toxicity of local anaesthetics is due to the speed of rise of blood concentration. Therefore a rapid injection of a small volume may cause toxicity.

ESSENTIAL PRECAUTIONS

- Secure intravenous access before injection of any dose that may cause toxic effects.
- Always have adequate resuscitation equipment and drugs available before starting to inject.

TREATMENT OF TOXICITY

If a patient you are attending shows any signs or symptoms of toxicity during injection of local anaesthetic stop the injection and assess the patient.

Treatment is based on the A B C D of Basic Life Support Call for help while treating the patient:

A. Ensure an adequate airway; give oxygen in high concentration if available.

B. Ensure that the patient is breathing adequately. Ventilate the patient with a self-inflating bag if there is inadequate spontaneous respiration. Intubation may be required if the patient is unconscious and unable to maintain an airway.

C. Treat circulatory failure with intravenous fluids and vasopressors such as ephedrine (10mg boluses) if hypotension occurs. Adrenaline may be used cautiously intravenously in boluses of 0.5-1ml of 1:10,000 (1mg in 10ml) if ephedrine is either not available or not effective in correcting the hypotension. Treat arrhythmias.Start chest compressions if cardiac arrest occurs.

D. Drugs to stop fitting such as Diazepam 0.2-0.4mg/kg intravenously slowly over 5 minutes repeated after

10 minutes if required, or 2.5mg-10 mg rectally. Thiopentone 1-4 mg/kg intravenously may also be used in theatre.Observe the patient closely after any reaction.

Treatment of local anaesthetic toxicity is likely to have a good outcome if toxicity is recognised and basic resuscitation is started early. Monitor patients closely when using local anaesthetics.

If a reaction occurs:

- Prevent hypoxia, which will cause brain damage and make fitting or arrhythmias more difficult to control.
- Ensure that hypotension and arrhythmias are treated early.
- Ensure that fits are adequately treated.
- Most reactions are short-lived if the above advice is followed.

Although there are a wide range of recognised poisons and toxic substances, it is important to remember that some of the most well-known poisonous substances are in fact drugs that are presently used worldwide to treat various symptoms and cure diseases. Due to the complex nature of the human body, it is practically guaranteed that all useful drugs will produce unwanted effects, but some can produce positively dangerous effects. Toxicity refers to these sometimes-deadly effects: though all drugs have adverse effects at normal doses, it is only the dangerous ones, which are referred to as toxic.

Drugs may have toxic effects on various organs, with some potentially useful drugs being abandoned due to toxic effects, and overdoses of some existing drugs leading to death. Doctors are often placed in difficult positions while trying to decide whether the benefits of a drug outweigh its risks and must be forever vigilant in looking for unknown effects.

A HISTORY OF DRUG TOXICITY

One of the first major incidences of drug toxicity in the 20th Century was in the USA in 1937 and involved a drug known as sulphanilamide, an anti-bacterial agent used to treat streptococcal infections. The drug had been used safely for quite some time in its tablet and powder forms, but it emerged that there was a call for the drug to be available in liquid form. This led to the production of an 'elixir of sulphanilamide', which the pharmacists in the USA found would dissolve in diethylene glycol. With the promise of increased profits, the elixir was immediately mass-produced without testing and sent all over the country. In the space of one month, over a hundred patients died, but it was not the drug that killed them. The solvent, diethylene glycol, is rather similar to common antifreeze (ethylene glycol) and is a deadly poison, but this fact had somehow been ignored.

Thalidomide

Although this sort of occurrence is now avoided by careful testing of new formulae, other drugs can have toxic effects that are much more difficult to detect. One of the most famous examples of a drug prescribed to the public and then hastily withdrawn is thalidomide. Introduced in 1957 in West Germany to treat morning sickness in pregnant women, thalidomide had not been tested thoroughly enough to discover its teratogenic effects. It was only when the pregnant women who had been treated with the drug started giving birth to deformed children with shortened limbs and no external ears that thalidomide was withdrawn.

By then, the drug was being used throughout Europe and had damaged 15,000 foetuses: 12,000 were deformed and half went on to die within the first year of life. Although the use of thalidomide in pregnant women is now banned, a new use for the drug has been found in patients with multiple myeloma and leprosy, and also has applications in Behcet's syndrome, AIDS and tuberculosis. This is partly due to the fact that the drug has been refined so that it only includes the (+) isomer, which has beneficial effects, and not the (-) isomer, which caused the many deformities associated with the drug.

Many More Examples

Although pharmacologists now had the experience provided by the thalidomide tragedy, it still proved difficult to detect the harmful side effects of many drugs. In the four decades that followed, many discoveries were made, all of them after the drug in question was released:

- The contraceptive pill, which in the 1960s contained 75mg of oestrogen, was found to increase the rate of thrombosis to a much higher rate than that found in pregnant women. Using only a third of the quantity of oestrogen has since reduced the risk.
- Clioquinol, an anti-infective agent, was found to cause SMON (subacute myelo-opticoneuropathy), a condition affecting the spinal cord and optic nerve, leading to disturbances of vision and movement as well as mental disorders.
- It was discovered that phenytoin, an anti-epileptic drug, could cause decalcification of bones, leading to rickets in children and osteomalacia in adults. This is due to the drug increasing the production of a hormone that decalcifies bone.
- During the 1970s, the beta-blocker practolol was found to cause an immune reaction known as oculomucocutaneous syndrome, causing inflammation of the eyes, mucous membrane and skin.
- In 1982, benoxaprofen, an NSAID (non-steroidal anti-inflammatory drug) used to treat arthritis, was released onto the market after tests on rhesus monkeys showed that it was safe. Sixty deaths and 3,500 serious illnesses occurred in the UK alone before the drug was identified as damaging the liver and also causing inflammation when patients were exposed to sunlight.
- In 1984, it was discovered that fenclofenac, another NSAID used to treat arthritis and inflammation, also caused severe liver damage. It had been released to the public after passing tests on ten different species of animal.
- Ten years later, it was found that the NSAID phenylbutazone caused liver and bone marrow damage as well as inflammation on exposure to

sunlight. The effects on bone marrow were only discovered in lab cultures of human cells after many cases of aplastic anaemia, a disease where the bone marrow fails to produce sufficient blood cells.

- During the late 1990s, both pemoline, a treatment for ADHD (attention deficit hyperactivity disorder), and troglitazone, a drug used to regulate type-2 diabetes, were both shown to damage the liver.

All of these examples illustrate the difficulties in detecting harmful effects in advance of actually giving the drug in question to humans. Although the governments of most countries require drugs to be tested on animals before use on humans, these tests can be deceptive. Harmful effects seen in humans are often not seen in rodents, and it is also conceivable that drugs, which are perfectly harmless to humans, could have adverse effects on other species. There are also laws limiting the numbers of animals that can be used, meaning that rare toxic effects that only affect animals with certain conditions are unlikely to be seen until the drug is released for use by humans. The advent of good clinical practice in the wake of the thalidomide tragedy has helped to reduce the effects of these problems, but they are still a hazard that needs careful monitoring.

Different Types of Drug Toxicity

Just as drugs come in many varieties and have varying beneficial and adverse effects, so there are different ways in which they can cause potentially toxic effects. Often it is not the drug itself that causes the trouble, but a breakdown product or 'metabolite', which remains active and has unwanted actions. There are even more complex situations than this, such as in the case of thalidomide, where the drug taken is broken down by the body into two components. One of these acts to reduce morning sickness, while the other causes malformation of the foetus.

In some cases, drugs are in fact chosen for their toxic nature when being used to treat malignant tumours and metastases. In these cases, the cytotoxic and mutagenic properties of these drugs are selected to kill mainly the cancer cells, but can also have negative effects on normal cells.

Cytotoxicity

This is the simplest form of drug toxicity, where the drug or an active metabolite causes serious damage to the cells. Often, the cells of a specific organ are affected, causing a potentially fatal loss of function of the liver or kidney, damage to the eyes or ears, or abnormal clotting of the blood. Other organs can also be affected: aspirin, for example, is known to cause stomach ulcers in some patients, especially after an overdose.

Another well-known cytotoxic drug is paracetamol, also known as acetaminophen, which can fatally damage the liver when taken in high doses. This happens because the body into various metabolites breaks down paracetamol, one of which is NAPQI. This is harmful to the liver, but is normally inactivated by glutathione. However, the body only has limited reserves of glutathione, and an overdose of paracetamol will produce enough NAPQI to deplete these reserves, allowing the NAPQI to destroy the liver cells freely. Paracetamol overdose is treated using methionine or N-acetylcysteine, both of which raise the levels of glutathione and thus inactivate the harmful NAPQI. It is now possible to determine quickly the level of paracetamol in the blood by using anenzyme immunoassay, allowing those with a known toxic level to be treated promptly.

Carcinogenicity

Some drugs can cause damage or mutations to DNA, leading to changes in cell metabolism and the promotion of tumours. Carcinogenic drugs may also activate oncogenes-versions of genes, which promote disregulation of cell processes-leading to the transformation of normal cells into cancerous cells. An example of a drug with carcinogenic properties is stilboestrol, an oestrogen analogue that increases the probability of developing endometrial cancer or breast cancer.

Mutagenicity

Some drugs can cause permanent changes to the DNA of germ cells-egg cells and sperm cells-leading to mutations, which are inherited by a patient's children. One example is nitrogen mustard, a compound similar to mustard gas, which is used to destroy cancer cells in chemotherapy. The drug destroys cancer cells by linking their DNA strands together with a nitrogen atom, but can also link the DNA of germ cells, leading to deletions of DNA bases, thereby causing mutations. For this reason, known mutagens are usually only used deliberately to kill cancer cells.

Teratogenicity

As in the case of thalidomide, some drugs can cause defects in the development of the foetus, leading to gross abnormalities of the baby: the term for this kind of toxicity, teratogenicity, derives from the Greek *teratos*, meaning 'monster'. The type of abnormality seen depends on the stage at which the drug is taken, as different organs develop at different times during pregnancy. For instance, thalidomide taken after 21 days of pregnancy will often lead to lack of external ears and paralysis of the cranial nerves, whereas thalidomide taken after 27 days will lead to limb malformation. This latter deformity is known as phocomelia, meaning 'seal-like limbs'.

Drug Allergy

Allergies are a well-known problem, with the offending substances ranging from peanuts to pollen. Drug allergies work in the same way, with the body developing antibodies to a drug, causing the immune system to overreact, leading to the usual range of symptoms including itching, wheezing and rashes. However, when the drug causes an anaphylactic shock, where the patient's immune system responds strongly enough to cause difficulties in breathing, low blood pressure, inflammation and even heart failure, the drug is noted as toxic to that individual, and cannot be given to them again. Penicillin is a good example of a drug that can cause allergic reactions and anaphylaxis, with a breakdown product of the drug causing an immune reaction.

Measuring Toxicity

A simple yet crude measure of the toxicity of a drug is the LD_{50}, which is the dose of a drug which kills 50% of treated animals within a specified short period of time. This quantity varies widely from ten grams per kilogram of body weight for alcohol, to just 0.00001 milligrams per kilogram for the botulinus toxin produced by the botulinium bacterium. Various other animal tests are used, and drugs are also added to colonies of bacteria and cultures of human cells to determine their carcinogenic and mutagenic effects.

Avoiding Unwanted Toxicity

Unwanted toxic effects can be avoided by banning dangerous drugs, or by strictly controlling their uses. Drugs must go through both rigorous animal testing and small-scale trials on humans to detect unexpected effects, with some dangerous drugs only being dropped after humans have reacted to them.

It is now common practice for pharmaceutical companies to advise doctors not to prescribe their drugs to pregnant women and up to half the drugs listed in the British National Formulary come with such warnings. Women of child-bearing age are often advised to use contraception while taking potentially teratogenic drugs.

Meanwhile, drugs that are potentially fatal after an overdose are sold in restricted amounts in blister packs, making it difficult to take enough pills for the drug to have toxic effects. Childproof containers have also been developed and clear labelling is used to warn patients of risks. Doctors are now very careful when prescribing drugs, which may interact with one another to cause increased effects, and they are also careful to keep up-to-date with the latest toxicity reports in journals. Previously unknown adverse effects of drugs are reported to the authorities using a standard procedure.

Treating Unwanted Toxicity

Treatment of toxicity refers to the treatment of overdoses, where the drug is likely to cause harm if its effects are not quickly countered.

The first line of defence is to remove the drug from the patient before it is fully absorbed, and techniques include:

- Irrigation to remove drugs applied to the eyes or skin.
- Gastric lavage, where the stomach is washed out and drained using tubes.
- Activated charcoal, which is swallowed and soaks up the drug from the gut.
- Ipecac syrup, which causes vomiting in order to empty the stomach.
- Cathartics, laxatives which purge the drug from the gut.

A second line of defence involves the removal of the drug from the bloodstream by various methods, including:

- Changing urine pH to increase excretion of the drug into the urine.
- Forced diuresis, where drugs are given to increase urine production.
- Haemodialysis, where the blood is passed through a machine to remove the drug.
- Exchange transfusion to replace the patient's drug-filled blood with fresh blood.

A vitally important line of defence is the administration of an antidote, which either removes the harmful substance from the blood or counters its effects.

Pralidoxine is used as an antidote to nerve gas, and ethanol is given to counter methanol, while hydroxycobalamin is used to treat cyanide poisoning, although it must be administered quickly.

An example of an antidote with an opposing effect is atropine, which in large doses causes muscle paralysis: this is used to oppose the effects of an overdose of anti-cholinesterases, drugs which increase the firing of nerves attached to muscles.

And Finally

Having read the above, it is very easy to worry about the possibility of any drug having potentially harmful effects. It is of course true that many drugs can be dangerous at high doses, but it is for this reason that doctors only prescribe small quantities of drugs.

Drugs, which can have carcinogenic effects usually only, cause cancer in a minority of patients and are used because their benefits vastly outweigh the risks. Meanwhile, potentially teratogenic drugs are not prescribed to pregnant women and mutagenic drugs are usually only used in chemotherapy, where the mutagenic effect is required to kill off cancer cells. All in all, doctors are provided with a vast quantity of information about harmful effects in the British National Formulary and have been trained to make careful decisions about which drugs to prescribe to patients.

A new computational method that searches an enormous database of protein structures could allow researchers to predict a drug's potential side effects without breaking out a single test tube. The technique, developed by researchers at the University of California, San Diego (UCSD), could also be applied to existing drugs to explain known side effects or to identify additional uses.

"The approach that they're taking is different and certainly way more computationally intensive than other approaches,"a pharmacologist at the University of North Carolina, Chapel Hill, who was not involved with the work.

Drugs work by latching onto very specific receptor sites on protein targets, much like a key fitting into a lock. If a drug also happens to latch onto another, unintended target, it may produce unwanted side effects. The new method analyses the shape of the intended lock and then combs through the structures of other proteins to look for similarly shaped locks. If such a lock is found, an algorithm rigorously determines whether the drug fits snugly into it. If it does, the technique has probably identified what's called an "off-target."

The researchers, led by UCSD pharmacologist Philip Bourne, didn't stop there. Once they had identified an off-target, they worked backward, looking for other drugs known to bind to it and then using their technique to determine whether those drugs could also bind to the original target. If so, the researchers had further evidence that the two proteins—the intended target and the unintended target—had highly similar drug receptor sites. Then, by assessing the biological function of the off-target, they could postulate possible side effects of the drug being tested.

DRUG SIDE EFFECTS

All medicines may cause side effects, but many people have no, or minor, side effects.Check with your doctor if any of these most COMMON side effects persist or become bothersome when using Methotrexate: Acne; chills and fever; dizziness; flushing; general body discomfort; hair loss; headache; infertility; irregular periods; itching; loss of appetite; lowered resistance to infection; miscarriage; nausea; sensitivity to sunlight; sore throat; speech impairment; stomach pain; swelling of the breast; unusual tiredness; vaginal discharge; vomiting. Seek medical attention right away if any of these SEVERE side effects occur when using Methotrexate: Severe allergic reactions (rash; hives; itching; difficulty breathing; tightness in the chest; swelling of the mouth, face, lips, or tongue); black or bloody stools; blood in the urine; bone pain; calf pain/swelling; change in amount of urine; chest pain; confusion; dark urine; diarrhea; dry cough; enlarged glands; fatigue; fever or chills; inflammation of the pancreas (stomach tenderness, nausea, vomiting, fever, increased pulse rate); irregular heartbeat; mental changes; mouth sores; muscle weakness; persistent sore throat; red, swollen, or blistered skin; seizures; serious infection (herpes, hepatitis, blood infection); trouble breathing; unusual bleeding or bruising; unusual pain and discoloration of the skin; vision changes; vomit that looks like coffee grounds; yellowing of skin or eyes.

Tablets

All medicines may cause side effects, but many people have no, or minor, side effects.Check with your doctor if any of these most COMMON side effects persist or become bothersome when using Methotrexate Tablets: Dizziness; headache; loss of appetite; mild hair loss; nausea; stomach pain or upset; tiredness; vomiting.Seek medical attention right away if any of these SEVERE side effects occur when using Methotrexate Tablets:

Severe allergic reactions (rash; hives; itching; difficulty breathing; tightness in the chest; swelling of the mouth, face, lips, or tongue); black, tarry stools; blood in the urine; calf or leg pain or swelling; change in the amount of urine; chest pain; coma; confusion; coughing up blood; dark urine; diarrhea; difficult or painful urination; dry cough; fever, chills, or sore throat; menstrual changes; mental or mood changes; mouth sores; one-sided weakness; red, swollen, peeling, or blistered skin; seizures; severe or persistent nausea or vomiting; shortness of breath; speech changes; unusual bleeding or bruising; unusual pain and discoloration of the skin; unusual tiredness or weakness; vision changes or vision loss; yellowing of skin or eyes.

Methotrexate

In General, The Incidence And Severity Of Acute Side Effects Are Related To Dose And Frequency Of Administration. The Most Serious Reactions Are Discussed Above Under Organ System Toxicity In The Precaution Section. That Section Should Also Be Consulted When Looking For Information About Adverse Reactions With Methotrexate.

The most frequently reported adverse reactions include ulcerative stomatitis, leukopenia, nausea, and abdominal distress. Other frequently reported adverse effects are malaise, undue fatigue, chills and fever, dizziness and decreased resistance to infection.

Other adverse reactions that have been reported with Methotrexate are listed below by organ system. In the oncology setting, concomitant treatment and the underlying disease make specific attribution of a reaction to Methotrexate difficult.

- *Alimentary System:* Gingivitis, pharyngitis, stomatitis, anorexia, nausea, vomiting, diarrhea, hematemesis, melena, gastrointestinal ulceration and bleeding, enteritis, pancreatitis.

- *Blood and Lymphatic System Disorders:* Suppressed hematopoiesis causing anemia, aplastic anemia, pancytopenia, leukopenia, neutropenia, and/or thrombocytopenia, lymphadenopathy and lymphoproliferative disorders.
- *Cardiovascular:* Pericarditis, pericardial effusion, hypotension, and thromboembolic events (including arterial thrombosis, cerebral thrombosis, deep vein thrombosis, retinal vein thrombosis, thrombophlebitis, and pulmonary embolus).
- *Central Nervous System:* Headaches, drowsiness, blurred vision, transient blindness, speech impairment including dysarthria and aphasia, hemiparesis, paresis and convulsions have also occurred following administration of Methotrexate. Following low doses, there have been occasional reports of transient subtle cognitive dysfunction, mood alteration, unusual cranial sensations, leukoencephalopathy, or encephalopathy.
- *Hepatobiliary Disorders:* Hepatotoxicity, acute hepatitis, chronic fibrosis and cirrhosis, decrease in serum albumin, liver enzyme elevations.
- *Infection:* There have been case reports of sometimes fatal opportunistic infections in patients receiving Methotrexate therapy for neoplastic and non-neoplastic diseases. Pneumocystis carinii pneumonia was the most common opportunistic infection. There have also been reports of infections, pneumonia, sepsis, nocardiosis, histoplasmosis, cryptococcosis, Herpes zoster, H. simplex hepatitis, and disseminated H. simplex.
- *Musculoskeletal System:* Stress fracture.
- *Ophthalmic:* Conjunctivitis, serious visual changes of unknown etiology.
- *Pulmonary System:* Respiratory fibrosis, respiratory failure, interstitial pneumonitis deaths have been reported, and chronic interstitial obstructive pulmonary disease has occasionally occurred.

- *Skin:* Erythematous rashes, pruritus, urticaria, photosensitivity, pigmentary changes, alopecia, ecchymosis, telangiectasia, acne, furunculosis, erythema multiforme, toxic epidermal necrolysis, Stevens-Johnson Syndrome, skin necrosis, skin ulceration, and exfoliative dermatitis.
- *Urogenital System:* Severe nephropathy or renal failure, azotemia, cystitis, hematuria; defective oogenesis or spermatogenesis, transient oligospermia, menstrual dysfunction, vaginal discharge, and gynecomastia; infertility, abortion, fetal defects.

Other rarer reactions related to or attributed to the use of Methotrexate such as nodulosis, vasculitis, arthralgia/myalgia, loss of libido/impotence, diabetes, osteoporosis, sudden death, reversible lymphomas, tumor lysis syndrome, soft tissue necrosis and osteonecrosis. Anaphylactoid reactions have been reported.

Double-Blind Rheumatoid Arthritis Studies

The approximate incidences of Methotrexate attributed (*i.e.,* placebo rate subtracted) adverse reactions in 12 to 18 week double-blind studies of patients (n = 128) with rheumatoid arthritis treated with low-dose oral (7.5 to 15 mg/week) pulse Methotrexate are listed below.

Virtually all of these patients were on concomitant nonsteroidal anti-inflammatory drugs and some were also taking low dosages of corticosteroids. Hepatic histology was not examined in these short-term studies. Incidence Greater Than 10%: Elevated liver function tests 15%, nausea/vomiting 10%. Incidence 3% to 10%: Stomatitis, thrombocytopenia, (platelet count less than 100,000/mm3). Incidence 1% to 3%: Rash/pruritus/dermatitis, diarrhea, alopecia, leukopenia (WBC less than 3000/mm3), pancytopenia, dizziness.

Two other controlled trials of patients (n = 680) with Rheumatoid Arthritis on 7.5 mg to 15 mg/wk oral doses showed an incidence of interstitial pneumonitis of 1%. Other less common reactions included decreased hematocrit, headache, upper respiratory infection, anorexia, arthralgias, chest pain, coughing, dysuria, eye discomfort, epistaxis, fever, infection, sweating, tinnitus, and vaginal discharge.

Adverse Reactions in Psoriasis

There are no recent placebo-controlled trials in patients with psoriasis. There are two literature reports describing large series (n = 204, 248) of psoriasis patients treated with Methotrexate. Dosages ranged up to 25 mg per week and treatment was administered for up to four years. With the exception of alopecia, photosensitivity, and "burning of skin lesions" (each 3% to 10%), the adverse reaction rates in these reports were very similar to those in the rheumatoid arthritis studies. Rarely, painful plaque erosions may appear.

Adverse Reactions in JRA Studies

The approximate incidences of adverse reactions reported in pediatric patients with JRA treated with oral, weekly doses of Methotrexate (5 to 20 mg/m2/wk or 0.1 to 0.65 mg/kg/wk) were as follows (virtually all patients were receiving concomitant nonsteroidal anti-inflammatory drugs, and some also were taking low doses of corticosteroids): elevated liver function tests, 14%; gastrointestinal reactions (*e.g.*, nausea, vomiting, diarrhea), 11%; stomatitis, 2%; leukopenia, 2%; headache, 1.2%; alopecia, 0.5%; dizziness, 0.2%; and rash, 0.2%. Although there is experience with dosing up to 30 mg/m2/wk in JRA, the published data for doses above 20 mg/m2/wk are too limited to provide reliable estimates of adverse reaction rates.

SIDE EFFECTS BY BODY SYSTEM

Hematologic

Hematologic side effects have included myelosuppression, which is one of the primary toxic effects of methotrexate. Methotrexate suppressed hematopoiesis has been reported to have caused anemia, aplastic anemia, pancytopenia, leukopenia, neutropenia, thrombocytopenia, lymphadenopathy, and lymphoproliferative disorders including reversible hypogammaglobulinemia (which has been reported rarely).

Preexisting myelosuppression or low hematologic cell counts are contraindications to the use of this drug, particularly in patients with rheumatoid arthritis or psoriasis. Close monitoring of the CBC is mandatory. Profound count nadirs may require therapy discontinuation, at least temporarily. Folate therapy and/or leucovorin rescue may be preventive or palliative. Treated patients who become febrile should be assumed to have neutropenia until proven otherwise.

Cytopenia occurs in 5% to 25% of patients with rheumatoid arthritis (RA) who receive long-term therapy. Risk factors include renal dysfunction, preexisting folate deficiency, increased mean corpuscular volume value, advanced age, concomitant use of other anti-folate medications (such as trimethoprim-sulfamethoxazole), and possibly hypoalbuminemia, concomitant infection, history of bone marrow injury, surgery, and concurrent use of NSAIDs or probenecid. Pancytopenia is rarely observed in patients with rheumatoid arthritis. Bone marrow recovery typically occurs within two weeks after the withdrawal of MTX.

Gastrointestinal

Gastrointestinal side effects are usually controlled by folate supplementation (1 to 5 mg orally, given 4 hours before MTX or 1 mg orally once a day if on low dose therapy, as in rheumatoid arthritis), dosage reduction, dividing the dose over a 12 to 24 hour period, withholding the drug, or giving it parenterally.

Extremely rare cases of colitis and toxic megacolon have been associated with the use of MTX. Gastrointestinal side effects, especially with high-dose administration, may be expected. Serious nausea, vomiting, diarrhea, or stomatitis (10% to 80% of patients followed in long-term studies) may result in symptomatic dehydration. Other frequently reported GI side effects, particularly after high-dose therapy, include gingivitis, pharyngitis, stomatitis, anorexia, hematemesis, melena, gastrointestinal ulceration and bleeding, and enteritis.

GI symptoms are often eliminated by folate supplementation. Folate does not affect efficacy of MTX.

Hepatic

Hepatic side effects including hepatotoxicity, acute hepatitis, chronic fibrosis and cirrhosis, decrease in serum albumin, and liver enzyme elevations have been reported.

Methotrexate can cause acute elevations of liver function tests (elevated serum transaminases in 15% of patients with rheumatoid arthritis (RA) on low-dose therapy) or chronic hepatotoxicity (fibrosis and cirrhosis). The incidence of liver fibrosis and cirrhosis in patients with RA (low doses) averages 3% to 7% and 0.1%, respectively. Meta-analysis has revealed the incidence of progression of liver disease (worsening of 1 grade on the histologic classification of Roenigk) in patients with RA or psoriatic arthritis averages 27%, or 7% per gram of MTX (total dose) given. Chronic hepatotoxicity typically develops only after chronic use of higher doses (2 years or more of total doses of 1.5 grams or more), is more likely in patients who ingest ethanol, who are aged, who are obese, who have chronic renal insufficiency, or who have diabetes.

In general, the following serve as guidelines for patients with non-oncologic conditions:

- Screen all patients to be treated with complete liver blood tests (transaminases, albumin, alkaline phosphatase, bilirubin), hepatitis B and C serologic studies (some experts recommend viral serology only in patients with abnormal liver function tests), baseline serum creatinine and complete blood count (CBC);
- Baseline liver biopsy if the patient has preexisting liver disease, persistently abnormal baseline AST (aspartate aminotransferase), history of excessive alcohol consumption (greater than 3 drinks/day), or chronic hepatitis B or C infection;
- Monitor hepatic enzymes every 4 to 8 weeks (with the understanding that they are not necessarily predictive of fibrosis/cirrhosis), discontinuing MTX

if serum transaminase levels meet or exceed 2 to 3x baseline (some guidelines recommend withholding MTX for 1 to 2 weeks if significant abnormalities persist);

- Liver biopsy should be considered if significant hepatic enzyme abnormalities persist for 2 to 3 months. Liver biopsy should be performed if 6 of 12 monthly AST levels are above the upper limit of normal or serum albumin levels are less than normal in the setting of well controlled disease;
- If liver biopsy shows Roenigk grade I, II, or IIIA (mild fibrosis), MTX may be resumed with monitoring as described above; if liver biopsy show Roenigk grade IIIB (moderate fibrosis) or IV (cirrhosis), MTX should be discontinued;
- MTX should be discontinued in patients with significant liver abnormalities, as described above, who refuse liver biopsy;
- It is still debated whether post-therapy biopsy should be performed every 2 to 3 years or after each 1.5 to 2.0 grams cumulative dose given, regardless of liver chemistries.

Pre-biopsy coagulation studies and withholding of aspirin and other NSAIDs are recommended.

Immunologic

New and/or opportunistic infection can arise during or after therapy with methotrexate due to drug-induced immune suppression. Infections during MTX therapy may occur in up to 58% of patients during low-dose therapy (as in RA). Patients are at great risk after high-dose therapy. MTX should be held prior to and during elective surgery to minimize the risk of infectious complications. Limited data have shown that the use of low-dose MTX for patients with rheumatoid arthritis is associated with decreased CD8 and naive CD4 T lymphocytes after only 8 weeks of therapy.Methotrexate is usually contraindicated in patients with overt or laboratory evidence of immunodeficiency syndromes. Immunization may be ineffective when given during methotrexate therapy. Immunization with live virus vaccines is generally not recommended. There have been reports of disseminated vaccinia infections after smallpox immunization in patients receiving methotrexate therapy.

Immunologic side effects including case reports of sometimes-fatal opportunistic infections have been reported. Pneumocystis carinii pneumonia has been reported most frequently. Also of concern are infections associated with severe immunosuppression, such as disseminated herpes zoster, Listeria meningitis, Mycobacterium avium intracellulare pneumonia, and systemic fungal infections (cryptococcosis, nocardiosis, aspergillosis, and histoplasmosis).

Nervous system

Nervous system side effects include headaches, dizziness, drowsiness, blurred vision, subtle cognitive dysfunction, moodiness, tinnitus or unusual cranial sensations. After intrathecal administration, headache, back pain, fever, and even transient paraplegias have been reported. A case of acute chorea and a case of a woman who couldn't speak after receiving methotrexate by intrathecal administration have also been reported. Serious neurotoxicity has been associated with the use of high-dose MTX after intrathecal or intraventricular administration to patients who have undergone craniospinal irradiation, but has also been described in other patients who have received low-dose oral therapy.The development of serious neurotoxicity is more likely if MTX is given in high doses to patients who have undergone craniospinal irradiation.

Significant neurotoxicity can present as an acute stroke-like encephalopathy or a chronic leukoencephalopathy. Unique features of the former include its acute onset (seizures, confusion, hemiparesis, speech problems, loss of consciousness) and reversibility within days.The use of high-dose IV MTX has been associated with the development of chronic delayed leukoencephalopathy in patients with or without a history of craniospinal irradiation. A syndrome of subtle personality changes followed by a progressive dementia, focal seizures, pseudobulbar palsy, spastic quadriparesis, and stupor may begin several months after the initiation of therapy. Affected patients often improve after discontinuation of therapy.

Intrathecal or intraventricular administration of MTX has been associated with the acute onset of headache, nausea, vomiting, fever, back pain, dizziness, and meningismus. Acute symptoms usually resolve within one to two days. In some cases, subacute neurotoxicity, with myelopathy or encephalopathy, may occur within days to weeks. Subacute neurotoxicity is usually characterized by paresis/plegia, dementia, confusion, tremor, ataxia, irritability, and somnolence.

Delayed leukoencephalopathy, as described above, has also been associated with the use of intrathecal MTX. The condition can improve but can be progressive and fatal.Progressive dementia and leukoencephalopathy has been reported in at least one case after low-dose, oral administration of MTX for the treatment of rheumatoid arthritis.

Respiratory

Two patterns of pulmonary toxicity have been reported: a hypersensitivity reaction and a toxic reaction with diffuse alveolar damage and nonspecific lung injury. Risk factors are unknown, but may include male gender, cigarette smoking, and concomitant use of an NSAID.

The hypersensitivity reaction is characterized by interstitial pneumonitis, granuloma formation, and the development of bronchopneumonia. Some patients have asthmatic symptoms.

Interstitial pneumonitis may be associated with duration of treatment or weekly or cumulative dose. The prevalence of this complication ranges from 2.5% to 7.5% in patients with psoriatic or rheumatoid arthritis. One prospective study has shown that the cumulative dose of MTX is independently related to small, but significant, increases in residual volume and that MTX may cause mild degrees of air trapping after long-term therapy.

The differential diagnosis of MTX pneumonitis includes infectious pneumonia, hypersensitivity pneumonitis caused by other drugs and autoimmune alveolitis due to rheumatoid arthritis (RA). Pneumocystis carinii pneumonia (PCP) can resemble MTX pneumonitis, and bronchoalveolar lavage (BAL) is a useful tool to distinguish between the two. BAL is not necessarily helpful in distinguishing MTX pneumonitis from RA alveolitis, however, since an elevated lymphocyte count and an increased CD4/CD8 ratio are observed in both conditions.

Pulmonary function tests in affected patients reflect a restrictive ventilatory defect and decreased oxygen-diffusing capacities. A mild peripheral eosinophilia is often present.

Respiratory side effects including toxicity can occur at any dosage in 3% to 11% of patients, and can mimic infectious pneumonia. Interstitial pneumonitis has emerged as one of the most unpredictable and potentially life-threatening side effects from low-dose MTX therapy. It presents as fever, dry/nonproductive cough, dyspnea, hypoxemia, and/or epistaxis, usually with chest X-ray infiltration. Signs or symptoms of infection must be treated seriously due to the possibility of MTX-induced neutropenia.

Renal

Renal side effects include renal insufficiency, which is most commonly associated with high-dose MTX since, after these doses, the concentration of a major circulating metabolite, 7-OH MTX, can precipitate in the renal tubule. Concomitant use of other potentially nephrotoxic drugs, including NSAIDs, and preexisting renal insufficiency are risk factors. There is a higher risk of nephrotoxicity in patients with underlying renal dysfunction. Aggressive and adequate hydration and urinary alkalinization helps minimize the risk of MTX-induced nephropathy, cystitis, and hematuria. If monitoring reveals renal dysfunction, decreasing MTX doses or discontinuing the drug altogether may improve renal function.

General

General side effects, which have been reported frequently,

include malaise, fatigue, and chills. Less commonly reported effects include arthralgias and myalgias. Leucovorin is typically given to diminish toxicity and counteract the adverse effects of high-dose therapy.

There is an increased risk of methotrexate (MTX) toxicity in patients who have "third space" accumulations of fluid, such as pleural effusions. MTX accumulates in these spaces and becomes cleared from these spaces slowly. These "deep pools" of MTX markedly increase the risk of toxicity, particularly gastrointestinal toxicity (mucositis) and should be evacuated before therapy begins.

Genitourinary

Genitourinary side effects may seriously affect either sex. Women may experience menstrual dysfunction, vaginal discharge, abortion, or infertility. Decreased libido has been described in each sex. Defective oogenesis and spermatogenesis is usually transient.

Dermatologic

Cases of severe, sometimes fatal, dermatologic reactions, such as toxic epidermal necrolysis, Stevens-Johnson syndrome, exfoliative dermatitis, skin necrosis, and erythema multiforme, have been reported in children and adults, within days of oral, intramuscular, intravenous, or intrathecal methotrexate administration. Reactions have been observed after single or multiple, low, intermediate or high doses of methotrexate for the treatment of neoplastic and non-neoplastic diseases.

Alopecia typically resolves several months after discontinuation. Dermatologic side effects include erythematous rashes, desquamation, epidermal necrosis, pruritus, urticaria, photosensitivity, pigmentary changes, alopecia, ecchymosis, telangiectasia, acne, cutaneous vasculitis, furunculosis, and alopecia. Painful plaque erosions have rarely been reported (when MTX was used to treat psoriasis).

Musculoskeletal

Musculoskeletal side effects from low-dose MTX in

patients with rheumatoid arthritis include rare instances of accelerated nodulosis. This problem is more commonly associated with MTX than other second-line agents. The use of MTX may be associated with decreased bone density in a syndrome called "MTX osteopathy" (this potential complication requires further investigation). This syndrome may be especially important in postmenopausal women.

Hypersensitivity

Hypersensitivity side effects including anaphylaxis have been reported rarely.

Oncologic

Oncologic side effects include some evidence that methotrexate may be oncogenic, particularly with respect to the development of some lymphomas and leukemias.

Large retrospective studies have shown that hematologic malignancies are uncommon in patients with RA treated with disease-modifying antirheumatic drugs, including MTX. These studies have shown that there does not appear to be a relationship between the peak or cumulative dose of the duration of MTX therapy and the subsequent development of hematologic malignancy. The histologic types of hematologic malignancies seen in MTX-treated patients do not appear different from those seen in patients with RA treated with other disease-modifying antirheumatic drugs. Underlying rheumatoid arthritis (RA) or Sjogren's syndrome are independent risk factors for the development of non-Hodgkin's lymphoma.

Cardiovascular

Cardiovascular side effects including pericarditis, pericardial effusion, myocardial ischemia, hypotension and ventricular arrhythmias have rarely been associated with MTX. Thromboembolic events including arterial thrombosis, cerebral thrombosis, deep vein thrombosis, retinal vein thrombosis, thrombophlebitis, and pulmonary embolus have also been reported with methotrexate use. Chemical pleuritis secondary to MTX has been identified as the cause of chest pain in some patients after high-dose therapy.

Endocrine

Endocrinologic side effects have included gynecomastia associated with the use of low-dose MTX in patients with rheumatoid arthritis.

Ocular

Ocular side effects reported with methotrexate include conjunctivitis and serious visual changes of unknown origin.

6

Nanotechnology in Chemical Modification

PATHWAY

The evolutionarily conserved phenomenon RNA interference (RNAi), the process by which specific mRNAs are targeted for degradation by complementary short-interfering RNAs (siRNAs), has increasingly become a powerful tool for genetic analysis and is likely to become a potent therapeutic approach for gene silencing. Consequently, understanding the mechanism of RNAi has become critical for developing the most effective RNAi methodologies for both laboratory and clinical applications.

The general mechanism of RNAi involves the cleavage of double-stranded RNA (dsRNA) to short 21–23-nt siRNAs. This processing event is catalyzed by Dicer, a highly conserved, dsRNA-specific endonuclease that is a member of the RNase III family. Processing by Dicer results in siRNA duplexes that have 52 -phosphate and 32 -hydroxyl termini, and subsequently, these siRNAs are recognized by the RNA-induced silencing complex.

Active RISC complexes (RISC*) promote the unwinding of the siRNA through an ATP-dependent process, and the unwound antisense strand guides RISC* to the complementary mRNA.The targeted mRNA is then cleaved by RISC* at a single site that is defined with regard to where the 52 -end of the antisense strand is bound to the mRNA target sequence. For RNAi-mediated mRNA cleavage and degradation to be successful, 52 -phosphorylation of the antisense strand must occur, and the double helix of the antisense-target mRNA duplex must be in the A form.

One highlighted difference between mammalian RNAi and RNAi in other eukaryotes is the lack of an amplification system for long-term persistence of RNAi in mammalian cells. For example, in Drosophila, ~35 molecules of dsRNA can silence ~1000 copies of the targeted mRNA per cell and can persist over the course of many generations. In mammalian cells, RNAi only persists effectively for an average of ~66 h before the siRNA is likely diluted out over the course of several cell divisions. The amplification that is seen in flies and other lower eukaryotes can potentially be attributed to three factors. One is that the conversion of long trigger dsRNA to smaller 21–23-nt siRNAs by Dicer adds a degree of RNAi amplification, whereas in mammalian cells long trigger dsRNA invokes the interferon response that activates the protein kinase PKR.

This suggests that only siRNA transfections successfully trigger RNAi in mammalian cells without other side effects, and thus, no amplification would take place through the processing of longer RNAs. A second factor in amplification is the presence of RNA-dependent RNA polymerase (RdRP), which has been found in plants, worms, fungi, and flies. RdRP has been postulated to amplify target mRNA, through a random, degradative PCR model, into dsRNA, which can be targeted by Dicer. However, no RdRP homologs have been found in mammalian cells, and the 32 -OH that is required for RdRP-dependent degradative PCR is not required for RNAi in mammalian cells, indicating that PCR-based amplification likely does not occur in mammals.

A third factor in amplification may be the high enzymatic turnover rate of RISC* during the targeting and cleavage of mRNA, which may add a degree of amplification to RNAi induction in all eukaryotes, including mammals. However, as the persistence of RNAi occurs for only a short period of time, finding methods for increasing the longevity of siRNAs in human cells will be fundamental for applying RNAi to laboratory and therapeutic applications. To address this issue of siRNA stability for prolonging the duration of dsRNA-mediated gene silencing and to further dissect the mechanism of RNAi in human cells, various chemically modified nucleotides predicted to affect siRNA stability were incorporated into siRNAs to study whether specific modifications increased or decreased the efficacy and persistence of RNAi in vivo.

The most important of these modifications was to the 22 -OH of the ribonucleotide that distinguishes RNA from DNA and is required for the nucleophilic attack occurring during the hydrolysis of the RNA backbone, the reaction catalyzed by degradative RNases. Our results showed that the 22 -OH was not required for RNAi, indicating that structural rather than chemical properties of siRNA–mRNA duplexes were the key to inducing RNAi and that RISC* did not require the 22 -OH for ribonuclease activity. 22 -modified siRNAs also increased the persistence of RNAi in human cells. Modifications that stabilized base-pairing interactions were also incorporated into the antisense strand of siRNAs and were able to initiate RNAi, signifying that this class of chemical modifications could be used to increase the targeting efficiency of siRNAs for mRNA target sequences and for allele-specific inhibition of gene expression.

Other chemical modifications affected the formation of the major groove of the A-form helix of the antisense-siRNA–target-mRNA duplex, and potentially disrupted H-bonds or sterically hindered protein contacts, most probably preventing the RISC* complex from stably interacting with the dsRNA duplex. These modifications completely abolished RNAi, demonstrating that an intact major groove in the A-form helix and stable RNA–protein interactions were required for RNAi in human cells. Finally, previous observations of psorelan cross-linked siRNAs implied that unwinding of siRNA occurred from the 52 -end of the antisense strand and that complete unwinding may not be necessary for effective RNAi.By using mismatched or chemically modified nucleotides on either the 32 or 52 half of the antisense strand within the siRNA duplex, we have shown here that RNAi depended on the integrity of the 52, and not the 32, half of the siRNA duplex, as defined by the antisense strand. Altogether, these results gave insight into the essential biochemical properties of functional siRNAs and how specific changes in the siRNA structure can affect the efficacy of RNAi. Furthermore, these studies present new methodologies for improving the stability and utility of siRNAs for future RNAi applications.

BETA-BLOCKERS

Beta blockers, also known as beta-adrenergic blocking agents, are medications that reduce your blood pressure. Beta blockers work by blocking the effects of the hormone epinephrine, also known as adrenaline. When you take beta blockers, the heart beats more slowly and with less force, thereby reducing blood pressure. Beta blockers also help blood vessels open up to improve blood flow. Beta blockers have been called "the musicians underground drug." Often musicians form their opinions, and may risk their health, based on locker-room-type information.

Performance anxiety can be a deeply personal subject for musicians, and many are reluctant to discuss all the possible remedies. It is our intention to bring this subject into the open, and to provide accurate information to inform personal opinions and decisions.

BETA BLOCKERS (SUCH AS INDERAL)

Beta blockers block the receptors for the physical effects of a person's natural fight or flight response. They are not sedatives, and they can't help anxiety of a purely psychological nature. Beta receptors are found in a number of places in the body: heart, lung, arteries, brain and uterus, to name a few. Like a key in a lock, beta blockers chemically fit into beta receptors and prevent norepinephrine from binding to the receptors that cause the symptoms of the fight-or-flight response. The degree of these effects depends on the dose and the individual's sensitivity to the medication. Peak effect occurs in one to one and a half hours. Ideally, this could allow a performer to play at his or her best, without the distraction or interference of excessive fight or flight symptoms.

Blocking beta receptors can cause decreased heart rate; decreased force of heart contractions; bronchoconstriction (can cause asthma attacks in people with asthma); uterine contractions; decreased blood pressure; relief of migraines; and decreased tremor.

The beta receptors found in the different areas of the body are not all the same, thus different beta blockers may affect these areas differently. For instance, metoprolol (Lopressor) and atenolol (Tenormin) are beta-1 selective, which means they block only beta-1 receptors found primarily in the heart, but not the beta-2 receptors found in the lung and uterus. Thus, they can decrease blood pressure, heart rate and force of cardiac contraction, but are less likely to cause bronchoconstriction and uterine contractions. This selectivity is not absolute and depends on the dose.

Some beta blockers enter the brain better than others. Propranolol (Inderal) crosses the blood-brain barrier particularly well. This may be why propranolol causes more central nervous system side effects, such as hallucinations, nightmares, and depression, than the beta blockers that do not cross into the brain as easily.

Interestingly, the ability of beta blockers to help anxiety seems related only to their blockade of beta receptors outside of the brain. Beta blockers will not help the emotional symptoms of stage fright (for example, sleep problems or negative inner voices).

Everyone's body responds differently to beta blockers. For this reason it is important that each person experiment, under medical supervision, to find the ideal dosage for a performance situation. Similarly, it is important to take the beta blocker a few times and under different circumstances before a major performance to be sure the individual has no adverse reactions.

DOCTOR ADIVISE

Everyone, without exception, must obtain their beta

blockers from a physician. It is a violation of federal law in the country, and many other countries, to give prescription medications to someone other than the person for whom they were prescribed.

Beta blockers are prescription medications for good reason. There are several conditions, such as slow heart rate, bronchial asthma, and heart failure, which can cause serious medical problems, even death, when taking this medication.

In addition, you should know that some psychiatrists and psychologists feel that beta blockers ought only to be used in the context of psychological intervention, and only as a temporary measure.

PROBLEMS ARE BETA BLOCKER

One reference lists 29 different uses, including high blood pressure, angina, irregular heart rhythms, migraines, prevention of a second heart attack, tremors, alcohol withdrawal, anxiety and glaucoma. In general, a physician can prescribe medications for uses not approved by the Food and Drug Administration, so long as there is scientific literature to support the use. Usually the pharmaceutical company involved will only research and develop a few uses to submit to the FDA to get the medication approved for marketing.

Once the medication wins FDA approval, the pharmaceutical company is only allowed to advertise it for FDA-approved uses. Over time, however, researchers and physicians gain experience using the medication for uses not approved by the FDA. Beta blockers are not FDA approved for use in anxiety or stage fright, but these medications can be prescribed for this purpose because there is scientific literature to support this use.

SAFETY OF BETA BLOCKERS

Only your physician can tell you whether beta blockers are safe for you. Beta blockers can significantly worsen some medical conditions. As a general rule, beta blockers should not be used in asthmatic people. Beta blockers can worsen congestive heart failure, Raynaud's syndrome, and diabetes.

Musicians who take beta blockers often do so in small doses and only on special occasions, so they may experience no side effects at all from their use. However, many side effects have been reported from these medications, including rash, anaphylactic shock (sudden unconsciousness or death), slowed heart rate, low blood pressure, cold extremities, worsening of heart failure, fainting, rapid heart beats, dizziness, fatigue, headache, depression, sleep disturbances, nightmares, hallucinations, short term memory loss, high or low blood sugar, stomach ache, flatulence, constipation, nausea, diarrhea, dry mouth, vomiting, heartburn, bloating, impotence or decreased libido, difficulty urinating, bronchospasm, cough, wheezes, naal stuffiness, joint pain, and muscle cramps.

BETA BLOCKERS CAUSE ADDICTION

Beta blockers do not cause addiction—which means physical and psychological dependence on a substance that is beyond the user's control. But beta blockers, when used on a regular basis, can cause physical dependence.

Physical dependence is when the body becomes accustomed to having a medication in the body, thus over time, more medication is required to achieve the same effect. It also means that if the medication is suddenly stopped, the person can experience a rebound in the symptom the medication was being used to treat. Not all medications cause physical dependence. To develop physical dependence, the medication must be used on a chronic basis.

In the case of beta blockers, physical dependence is not usually an issue when they are used for stage fright, as they are used in low doses and are not taken on a chronic basis. When beta blockers are taken on a chronic basis, as for high blood pressure, suddenly stopping them can cause a severe, dangerous elevation in blood pressure that could lead to a stroke or heart attack.

Addiction causes the person who drinks alcohol, abuses a medication, or takes an illegal substance to exhibit substance-seeking behaviour, perhaps even committing criminal acts to obtain it.

What makes this issue confusing is that alcohol, narcotics, cocaine, and amphetamines can cause both physical dependence and addiction, and usually both must be treated at the same time. Beta blockers do cause physical dependence when used on a chronic basis, but it would be highly unusual for them to cause addiction.

BETA BLOCKERS HELP IN PERFORMANCE

The answer varies greatly among individuals. Obviously, if you have a medical condition that makes beta blockers dangerous for you, they will not help. Beta blockers don't make you play better by themselves; they just relieve physical problems resulting from the fight or flight response. On the other hand, some musicians feel that adrenalin helps their performing, giving them an edge that adds intensity to the performance.

Beta blockers have not been shown to directly improve a musician's emotional state, except to the extent that some musicians feel better when their physical problems are relieved. If your performance anxiety shows itself mainly in psychological ways (*e.g.* negative inner voices), beta blockers will not help you. The scientific studies and articles we looked at clearly show that beta blockers significantly reduce symptoms that can hinder some people's playing. The musicians in the studies said they felt better about their performance after taking beta blockers, and music critics consistently judged their performances to be better.

A note about dry mouth: Anecdotal reports show that some musicians who've tried beta blockers say they don't help dry mouth, or can even make it worse. In our research, however, we found a study that conclusively showed beta blockers to help dry mouth among brass players.

BETA BLOCKERS FOR PERFORMANCE ANXIETY

As you might expect, opinions vary widely. Here are what a few well-known musicians have said on the subject: Trevor Wye, flutist and teacher: After saying that beta blockers "should seriously be considered for important occasions," he writes: "Readers who may have some misgivings about the use of alcohol or drugs may care to reflect on the effect of nerves and strain on the body which can cause a great deal of damage over a period of time.

However, one should use caution by understanding the side effects of a given drug and by realizing that an addiction to drugs may occur and that this may therefore not be the ideal approach to the problem of stage fright.

DEALING WITH PERFORMANCE ANXIETY

Because the fight or flight response occurs only when we perceive danger, it can be avoided or minimized if we can convince ourselves there is nothing to fear. This is the goal of many non-medication approaches. The physical sensations of fear can also be managed in specific, non-medication ways. Cognitive therapy, the area of psychotherapy that holds that your understanding of a situation can change your emotional response to it, is often sought by people with stage fright. Hypnosis may also used. If you decide to seek a therapist, try to find one who is experienced in treating anxiety disorders, and especially one who has experience treating performers.

DIFFERENT CLASSES OF BETA-BLOCKERS AND SPECIFIC DRUGS

Beta-blockers that are used clinically can be divided into two classes:

- Non-selective blockers (block both $b_1$and $b_2$receptors), or
- Relatively selective b_1 blockers ("cardioselective" beta-blockers).

Some beta-blockers have additional mechanisms besides beta-blockade that contribute to their unique pharmacologic profile. The two classes of beta-blockers along with specific compounds are listed in the following table.

CLINICAL DEVELOPMENT

The key to the IRCT approach is our ability to combine clinical trials conducted by Encode, our wholly-owned clinical trials subsidiary, with our population data. Encode has enrolled over 3,000 patients in over 30 clinical trials. In an IRCT the trial cohort is enriched by using population-based epidemiological data as well as data on genetic markers, gene-expression patterns and serum biomarkers correlated with a given disease. This enables deCODE's clinical team to select for a trial patients who are susceptible to disease through the pathway targeted by the drug and are thus likely to respond.

Because of the much larger amount of information going into and coming out of an IRCT, we can test therapeutics for major indications more quickly and with significantly smaller cohorts than would normally be required. The results of an IRCT can then be applied to understand and predict the responsiveness of the patient population at large, information that can be used in further clinical development or to bring the drug to market.

As profit and regulatory compliance pressures increase, life science companies are desperately seeking ways to decrease the cost and manage the risk of their clinical trials. The *Clinical Development Strategy and Technology* service provides life science companies with the necessary insight to increase their operational focus and sharpen their study planning and execution skills.

This Health Industry Insights service provides a forward-looking analysis of IT technologies on the critical path and how they are being adopted all along the clinical development value chain. In addition to qualitative assessment, this service includes quantitative market sizing and forecasts to provide a complete picture of the processes and technologies that impact clinical development.

APPROACH

This service develops comprehensive data and unique analysis through focused topical surveys, secondary research, and insights from industry experts, practitioners, and vendors. To ensure relevance, Health Industry Insights' analysts work with subscribers to identify and prioritize specific topics to be covered in research reports.

TOPICS ADDRESSED

Throughout the year, this service will address the following topics:

- The convergence of the eClinical ecosystem
- Clinical performance management and metrics; Clinical data management, Clinical trial management
- Critical tools and systems, including EDC, ePRO, CTMS, CDMS, IVR/IXR
- Evolution towards adaptive trials
- Achieving innovation and optimization of clinical trials IT infrastructure
- Cost containment and risk assessment in clinical development
- eClinical technology landscape, including integration with electronic health record (EHR) and other healthcare IT systems
- Pharmacovigilance and drug safety
- Regulatory compliance issues and data integration standards: CDISC, HL7, ICH

KEY QUESTIONS ANSWERED

Our research addresses the following issues that are critical to your success:

- How can both the cost and the risk of clinical trials be managed while maximizing time to market and data quality?
- What are the trends in technologies for trial design, patient recruitment, and site monitoring, and how will these trends affect end users in both medical device companies and the pharmaceutical industry?
- How can regulatory compliance be achieved while staying on time and within budget?
- What are the key software and IT infrastructure issues in clinical development?
- What are the key issues and best practices in data collection, data validation, and reporting?
- What are the best practices and tools to use to optimize trial management through accelerating

patient and investigator recruitment, managing sites efficiently, and balancing the portfolio of trials?

- What are the best practices seen from companies leading the implementation of IT in clinical trials?
- How do I ensure that my company's use of IT in clinical trials compares favourably with that of my competitors?

WHO SHOULD SUBSCRIBE

The adoption of eClinical technologies is rapidly moving from a tactical, project-based technology acquisition process to a strategic, enterprise-wide investment methodology. As a result, the insights provided from the *Clinical Development Strategy and Technology* service are beneficial to a growing number of individuals in the life science community, including chief information officers (CIOs), eClinical technology vendors, clinical operations, IT executives, and clinical investigators.

RETURN ON RESEARCH INVESTMENT

Core Benefits for Technology Buyers:

- A clear view into the benefits, as well as the risks, of adopting eClinical technologies
- The ability to select best-in-class solutions, data interchange standards, and technology vendors
- Strategies to maximize your investment in existing technologies

Core Benefits for Technology Solution Providers:

- An assessment of the market opportunity for eClinical technology solutions
- Unbiased insight from key opinion leaders in the industry
- Increased market exposure and competitive intelligence

STANDARD SUBSCRIPTION DELIVERABLES

Health Industry Insights values building client relationships and strives to foster a partnership with each client through continuous analyst interaction; concise delivery of current, timely, and relevant research; as well as custom research when needed. Each advisory service delivers strategic research covering business and technology issues.

A summary of the basic service deliverables available to subscribers includes the following:

- Analyst inquiry time specific to the advisory service area
- 24 × 7 research access via healthindustry-insights.com
- Discounted admission to specific Health Industry Insights or IDC conferences
- Complimentary forum invitations

JOIN OUR CONVERSATIONS

Our Insights Communities are designed to enable technology buyers to engage with our industry analysts, share knowledge and best practices, and connect with colleagues.

To participate in our communities, members create a profile to:

- Participate in our discussions in our blogs, forums and polls.
- Learn about and share best practices, tips, tricks and tools related to business/technology.
- Network with colleagues and join groups within each industry.
- Provide feedback to IDC's industry analysts and hear what we're up to.

RELATED OFFERINGS

Related offerings to our industry-specific research practices include IT Spending Guides-segmented regional and country-level analysis by solution area; Go-To-Market Services (GMS)-a modular set of deliverables to assist clients in extending their market reach, validating their market strategy, and producing more responsive customer-centric marketing campaigns; Global IT Advisor (GITA)-a flexible, cross-disciplinary research service providing granularity in market information and analysis across a wide range of technology areas; and IT Executive Strategies (ITES)-a best-practices research service covering IT governance, portfolio management, vendor procurement, and more. In addition, Health Industry Insights experts are available for individualized consulting engagements to address specific challenges or to customise the results of our syndicated research offerings.

Four trends stand out: continued rapid growth; increased focus on innovation; differentiation of opportunities; and India's increased importance. Like bamboo, Asia's green shoots grow faster than those in the West. IMS Health predicts low single-digit pharma growth for EU and the US, but double-digit growth rates for APAC over the next five years.

With much of their incremental growth revenues coming from APAC, over the next decade, biopharma companies are shifting R&D spend into the region. This is not just to leverage the well-publicized lower costs and faster recruitment benefits, but is also about market access—as placing global trials in APAC speeds up product launches and product penetration.

For us in the CRO space, this is of course very good news. Business for CROs is predicted to grow even faster than sponsor company growth. Why? Because in addition to the shift of R&D spend to APAC, many sponsor companies simply do not have large organizations in the region. Engaging a CRO means sponsors can go straight to a more agile, and flexible outsourcing model without complications of managing legacy and headcount issues. Growth remains the key word for APAC. Global biopharma and CRO companies are increasingly observing their success in APAC as a surrogate measure of their global success. The companies now realise that they cannot succeed globally without being successful in APAC.

Innovation is becoming a buzz word for APAC. Some of this is traditional biopharma innovation generated in the labs—and the world will see more novel compounds appear from companies such as S-Bio in Singapore. The next decade will also see APAC biopharma companies replicating the process and enabling technology innovation, which is seen in the IT and electronic industries. This innovation will act synergistically with Asian advantages for drug development.

A great example is the drug development enabling website, goBalto.com. This site (incidentally set up by a Korean entrepreneur Jae Chung) will enable Next Generation virtual drug development companies to more easily tap into the burgeoning talent in APAC. Imagine the formation of an Asia-enabled drug development machine, which leverages Asia's lower costs and faster project cycle times, to get products much faster and cheaper than in the West. This has the potential to change the global biopharma operating model in the same way that Japanese car manufacturers transformed the auto industry.

For global CROs, there's tremendous opportunity to facilitate innovation in APAC by providing a bridge between East and the West. As the region matures, greater differentiation is becoming apparent. I have always referred to the APAC region as a portfolio of countries, each with different opportunities. For example, the fast, efficient and predictable regulatory environment in Singapore allowing early site initiation, massive population in India and China providing rapid enrolment and Taiwan's pedigree in oncology studies. Now, there is more clarity about what these differences are, and experienced CROs are better able to allocate development projects across different locations to optimize the mix of benefits and minimize downsides.

Expect this trend to continue. Singapore provides a good example. Recognizing the limitations of its relatively small population, it is becoming a regional center-of-excellence in early-phase studies that require small subject numbers but a fast and predictable regulatory environment. Another 'small' populated country, which is building on an already strong reputation in the early-phase space, is Australia. Expect Australia to increase its focus on Health Economics and Outcomes Research (HEOR) work, as there is a tremendous experience base, thanks to the strict requirements of their regulatory and reimbursement environment.

CROs will also become more differentiated. People in APAC still have difficulty in explaining the differences between various CROs. However, the industry's rapid growth in the region will provide real opportunities to build differentiated CROs in the region. CROs will be differentiated on the basis of service offering and leveraging across centers-of-excellence in different locations in APAC. For example, biometrics capability in India and China, regulatory and HEOR consulting in Australia and market access insight in China and Korea. Differentiation also will occur with certain product related opportunities. APAC has become the center-of-excellence for oncology studies as highlighted in a recent research article. Biosimilars are an area of real opportunity for APAC with companies such as Celltrion in Korea making global impact in this growing sector. The final trend that is increasingly prominent is India's emergence as the hot growth area in drug development. Volumes have been written and many conference-hours spent discussing India's clinical research capabilities, yet the story is still unfolding. India continues to offer new, unanticipated and interesting opportunities. The original drivers of low cost and access to patients continue to be attractive. India is making its regulatory system more robust, stringent and effective, which will further increase the country's appeal.

Those traditional benefits will ensure India to grow rapidly in its drug development industry for many years. India is also the real source of innovation, particularly in how things can be done better and more efficiently. One such idea is reversing the usual talent deployment approach for successful delivery of international projects. Traditional deployment of talent involved placing project leaders in more mature clinical research markets, while India/APAC's participation remained limited to lower level functions in the value chain. However, working towards maturing local talent to higher level functions in a systematic way offers several advantages including enhanced efficiency, 24/7 leadership coverage for projects and a 'can do' mindset that really resonates well with sponsors. For example, local employees at Kendle India love the career opportunities this provides, with the result, we enjoy as one of the lowest turnover rates of the employees in the industry.

Macro level trends for India includes:

- Relatively stable economy and buoyant local biopharma companies and
- Well rounded progress of clinical development functions marked by addition of newer services, and as reflected in recent strategic deals initiated between top 10 multinational companies with Indian giants.

ROLE OF FOOD AND DRUG ADMINISTRATION

As a professional couple with one half in the world of human subjects' protection and the other half in the world of academic pulmonary medicine, we read the article by Miller and Shorr and the accompanying editorial by Perkins with great interest. The article, in the April 2002 issue of *CHEST*, elucidated the many challenges faced by all of us who conduct research involving human subjects. The authors have brought into focus the ethical challenges presented to the investigator, the institutional review board (IRB), and the sponsor. Although we agree with the authors regarding the problems that they have identified, we suggest that the responsibility for the problems is not limited to the three groups they have identified. We would like to point out some of the inherent and fundamental problems of industry-sponsored randomized clinical trials as they are currently conducted.

The current practice is that industry (with some input from expert consultants) designs clinical trials, the US Food and Drug Administration (FDA) reviews them, and then the sponsor then seeks sites for these trials. As Miller and Shorr point out, clinical trials have seen a tremendous change in recent times, moving away from academic medical centers and towards private practice settings.

The scrutiny of IRB applications for these trials is different given the context of an independent IRB relative to the academic IRB. There is a potential and very real problem of "IRB shopping" with the obvious benefit that sponsors can have clinical trials conducted, unopposed and without modification. IRBs, both local and independent, are under extreme pressure to review protocols quickly and without requesting modification. By contrast, there is a strong tradition and clear need for local control of IRBs so that their actions reflect the communities that they serve. However, Miller and Shorr have pointed out the universal nature of the ethical dilemma confronting the investigators, the IRB, and the sponsor. Obviously, these groups did not meet this ethical challenge in the mometasone trial. In this context, such review standards and, in fact, this level of review would be more appropriate at the level of the FDA. At the present time, the role that the FDA plays in the risk/benefit analysis for the subject is ambiguous. This realization leads to a variety of troubling questions. What was the role of the FDA in allowing a study such as the mometasone trial to proceed? In general, what kind of review does an investigational new drug application undergo? Do experts in the area of study review it? Does the FDA consider whether the protocol meets current guidelines for the treatment of the condition under study, or does the FDA see this as the responsibility of the sponsor or the local IRB?

Our local IRB is frequently told by sponsors that we are the only IRB that has issues with a particular protocol. These issues usually relate to our concerns with including a placebo arm, or with withdrawing or withholding proven treatment. Industry-sponsored clinical trials that do not pass ethical muster or put patients at unjustifiable risk should be stopped at the FDA review. They should be rewritten according to currently accepted guidelines for the treatment of the condition under study, and according to regulations meant to protect any vulnerable subjects who might be included.

At national IRB meetings last year (Public Responsibility in Medicine and Research), Robert Temple, MD, of the FDA suggested that the agency preferred to allow protocols to have broad inclusion and exclusion criteria so as not to limit multicenter trials. He saw narrowing of inclusion and exclusion criteria to make a study safe as the responsibility of the IRB. At the same IRB meeting, a representative of the pharmaceutical industry was quick to point out that academic IRBs were a problem for industry, and they would prefer that independent IRBs review all multicentered trials. The implications of this latter statement are obvious. However, the suggestion of the FDA that narrowing criteria should be done at the local IRB feeds this somewhat dysfunctional relationship. The federal government has now begun to realise that IRB shopping is a real phenomenon. Currently, proposed regulations to prohibit IRB shopping are available for public comment. However, prohibiting IRB shopping alone will not resolve the current problems involving the clinical trials industry that lead to conducting trials such as the one reviewed and analysed by Miller and Shorr.

Unquestionably, unethical clinical trials have been funded by sources other than industry, and have been approved by local IRBs, including academic IRBs. National Institutes of Health-funded studies have also failed the risk/benefit analysis. In this time of heightened public awareness of the shortcomings of the human subjects' protection process, we would argue that rigorous risk/benefit analysis should be applied before protocols reach the local IRBs. In this way, we can help prevent local conflicts of interest and ensure that research is conducted at a standard deserving of the public trust.

The establishment of FDA as it exists today grew out of a time early in the century when consumers were victimized by dishonest purveyors of fraudulent potions and compounds that were ineffective, dangerous, or both. A system of drug regulation was established in the United States that has served us well. Under this system, FDA reviews new drugs to assess their safety and efficacy. In addition, certain types of drugs must be prescribed and dispensed by only licensed health care professionals. The prescribing requirement is based on the principle that certain drugs have risks of such significance associated with them that they should be administered only under the supervision and recommendation of a "learned intermediary"-that is, a licensed practitioner with the education and training necessary to oversee the administration of potentially harmful drug products.

Similarly, these products may only be dispensed by a licensed professional that can help to assure proper dosing and administration, and can provide important information on the drug's use to patients. The types of unlawful conduct involving online drug sales that FDA has identified are similar to unlawful activities that occur in other sales contexts. Under the Federal Food, Drug, and Cosmetic Act, FDA has the legal authority to take action against:

- The importation, sale, or distribution of an adulterated or misbranded drug;

- The importation, sale, or distribution of an unapproved new drug;
- Illegal promotion of a drug;
- The sale or dispensing of a prescription drug without a valid prescription; and,
- Counterfeit drugs.

When the Internet is used for an illegal sale, FDA, working with the Department of Justice, must establish the same elements of a case, develop the same charges, and take the same actions as it would if another medium, such as a storefront or a magazine, had been used. FDA has investigated and referred cases for criminal prosecution and initiated civil enforcement actions against online sellers of drugs and other FDA-regulated products, particularly sellers of drugs not approved by the Agency. As will be described later, FDA has significantly expanded its enforcement activities during this past year with regard to online drug sales.

CODE OF FEDERAL REGULATIONS (CFR)

INTRODUCTION

Code Of Federal Regulations (CFR) The final regulations published in the Federal Register(daily published record of proposed rules, final rules, meeting notices, etc.) are collected in the CFR. The CFR is divided into 50 titles which represent broad areas subject to Federal regulations. The FDA's portion of the CFR interprets the Federal Food, Drug and Cosmetic Act and related statutes.

Section 21 of the CFR contains all regulations pertaining to food and drugs. The regulations document all actions of all drug sponsors that are required under Federal law. 21CFR Part 314-Applications for FDA Approval to Market a New Drug or an Antibiotic Drug.

CDER's Manual of Policies and Procedures (MaPPs). These documents are approved instructions for internal practices and procedures followed by CDER staff to help standardize the new drug review process and other activities. MaPPs define external activities as well. All MaPPs are available for the public to review to get a better understanding of office policies, definitions, staff responsibilities and procedures. MaPPS of particular interest to NDA applicants include: Review of the Same Supplemental Change to More than One NDA or ANDA in More Than One Review Division. Filing Review Issues. Action Packages for NDAs and Efficacy Supplements. Refusal to Accept Application for Filing From Applicants in Arrears.Drug Application Approval 50 Policy· Requesting and Accepting Non-Archivable Electronic Records for New Drug Applications.

Despite billions of dollars, decades of research, and an unparalleled level of international cooperation between research scientists and clinicians, cancer remains a major cause of death worldwide. Therapies have improved and many forms of cancer are now treatable, but the disease still kills over six million people throughout the world each year. For this reason, cancer research funding through the National Institute of Health (NIH) and the National Cancer Institute (NCI) represents one of the largest expenditures of the United States federal government. This focus on research has led to improved medical treatments for cancer as well as greater understanding of the molecular intricacies of the disease. Cancer is now considered not simply one disease but rather a multitude of independent disorders that can all result in malignant cellular growth. Unlike other diseases, there is no simple explanation, or even definition, for cancer, and so there is no simple way to treat it. The dream of a cure for cancer, the "magic bullet" to miraculously eliminate the disease, is now considered unrealistic in light of overwhelming complexities.

However, an interdisciplinary approach has emerged offering new hope for cancer therapies. Standard chemotherapy or radiation regimens are nonspecific in their targeting of cancer cells. In contrast, this new field of therapy reflects a growing awareness on the part of the oncology community that the most effective way to treat a tumor is to target the specific molecular mechanisms responsible for its uncontrolled growth. Research has shown that not all tumors are the same; molecular targeting, the attempt to tailor therapy to the specific abnormalities causing disease, has become one of the most promising areas in cancer biology research. In addition, as a specific application of molecular targeting, pharmaceutical research has combined the basic science of molecular biology and biochemistry with combinatorial chemistry and organic synthesis to create a new field: rational drug design.

Unfortunately, drug development takes time, and it is years, often decades, before a drug that shows promise in the lab will make it into the clinic. Rational drug design, because it narrows the focus on possible therapeutic agents, is a potential way to shorten this lengthy process. The inherent benefits of the approach are highlighted by one of the most exciting developments in cancer biology in recent years. On May 10, 2001, as a result of overwhelmingly positive clinical trials, the Food and Drug Administration (FDA) approved Novartis Pharmaceutical's Gleevec (imatinib mesylate) for use in the treatment of chronic myelogenous leukemia (CML). In a stunning validation of the value of rational drug design, Gleevec is the first of a wave of compounds with the potential to revolutionize the way we look at cancer.

THE TARGET-CHRONIC MYELOGENOUS LEUKEMIA

Leukemia is a disease of the white blood cells. While normal white blood cells, or leukocytes, are responsible for maintaining the immune system, the abnormal proliferation of such cells leads to a deadly form of cancer. Chronic myelogenous leukemia (CML) is characterized by the accumulation of immature granulocytes, a specific type of white blood cell. The disease progresses through three stages, and the specific stage of disease in a patient is crucial in planning treatment. In stage one, also known as chronic phase, there are few if any outward symptoms of disease; many CML patients are unexpectedly diagnosed as the result of blood work following a routine doctor's appointment. After a period of up to several years, chronic phase CML progresses to accelerated phase CML, which is characterized by an increasing ratio of malignant to normal white blood cells. The final stage of CML is known as blast crisis, and by this point almost a third of blood and bone marrow cells are cancerous. Due to their high numbers, some of these blast cells may metastasize and form tumors in other tissues including the bones or lymph nodes. Until recently, the most effective form of treatment for CML was the use of a compound called alpha interferon which was often followed by high dose radiation, chemotherapy, and a bone marrow transplant. However, as CML progresses, the efficacy of bone marrow transplants decreases sharply. Transplants done in the accelerated or blast crisis phase have a high rate of failure. Only 15% of patients transplanted in blast crisis are still living after five years (Leukemia Society).

Despite the grim prognosis for CML patients, there is one crucial molecular detail of the disease that renders it a promising target for rational drug design. The highly specific targeting that led to the laboratory development of Gleevec can be traced back to some of the early studies of leukemia. The first piece of the puzzle was uncovered in 1960 when it was discovered that nearly all patients with CML possess an abnormally small chromosome 22. This abnormality was nicknamed the Philadelphia chromosome and has since become known as CML's cytogenetic calling card. Shortly thereafter it was discovered that chromosome 9 was elongated by approximately the same length as the missing region of chromosome 22; these findings were eventually linked after molecular characterization confirmed a translocation between chromosomes 9 and 22. In the process of this exchange, breakpoint cluster region (bcr) proteins from chromosome 22 become fused with the c-abl oncogene from chromosome 9 thus creating a hybrid bcr-abl gene. A normal c-abl gene encodes a protein that is highly regulated; the functional enzyme is shuttled between the nucleus and cytoplasm of a cell so that it is only activated when needed. However, under the influence of bcr resulting from bcr-abl fusion, c-abl loses the checkpoints intended to hold it under control. The enzyme becomes a constitutively active cytoplasmic protein. Studies in mice have shown that the introduction of the bcr-abl fusion protein, regardless of the presence of the Philadelphia chromosome, is sufficient to experimentally induce the development of CML. The consequences of the bcr-abl abnormality are devastating because of the resulting c-abl enzyme that is produced. This unregulated enzyme, a tyrosine kinase, is the molecular cause of CML. Kinases are enzymes that transfer phosphate groups from ATP to particular residues, in this case tyrosines, of target proteins within the cell. Phosphorylation often serves as a molecular switch to activate or inactivate a range of proteins involved in such vital activities as cellular growth, development, and survival. For this reason, kinases are normally kept under tight control; a runaway kinase can promote the excessive proliferation associated with cancers including CML.

THE GLEEVEC APPLICATION

Since abnormal enzymes are implicated in a variety of human cancers, targeting the mutant function should have deleterious affects for the tumor while leaving normal tissue unaffected. As a result of studies highlighting the molecular mechanisms of CML, a search began for compounds to inhibit the runaway *bcr-abl* kinase. Compound screening at Novartis Pharmaceuticals in the early 1990's identified 2-phenylaminopyrimidines as a general class of kinase inhibitors, and subsequent chemical synthesis was used to design a series of compounds that should have more specific anti-tumor effects. One of the drugs in this class was called STI571.

In vitro studies using STI571 demonstrated that it was a highly effective and specific inhibitor of platelet-derived-growth-factor (PDGF) kinases, c-kit kinases, and c-abl kinases. STI571 acts by blocking the ATP-binding site of these enzymes. The functional form of the enzyme uses ATP as its source of phosphate groups, thus by blocking the ATP-binding site of the kinase, STI571 prevents the enzyme from becoming active. In the case of bcr-abl kinases, further experiments demonstrated that STI571 inhibited growth in cell lines carrying the bcr-abl mutation. Animal models produced data indicating an STI571 dose-dependent inhibition of tumor growth although tumors were not completely eliminated. On the basis of promising laboratory data, STI571 began clinical trials in 1998. The first trials were conducted in patients with chronic phase CML who had previously failed interferon therapy. As with any Phase I clinical trial, the main purpose of the study was to determine the overall safety and acceptable dose ranges of the drug. However, in the midst of safety evaluations, the physicians running the trial noticed that with elevated doses of STI571 an astounding 98% of the patients showed a complete hematologic response-the abnormal Philadelphia chromosome was still detectable, but the elevated levels of circulating white blood cells, the cause of CML symptoms, were gone.

Due to these astonishing results, more expansive Phase II trials were conducted in over 1000 patients representing a range of CML stages, 27 cancer centers, and six countries. In the 532 chronic phase patients, the results of the original trial were confirmed. Only 3% of patients experienced disease progression while a mere 2% did not complete the trial due to drug side effects. In addition, of the 86% of patients on long term STI571 treatment, 75% achieved either complete or major cytogenetic responses (Mauro and Druker). Not only did blood counts return to normal, but in many cases the molecular causative agent of disease, the Philadelphia chromosome, had become undetectable. These results were also observed in the 233 accelerated phase patients: 91% achieved a hematologic response, 41% had a cytogenetic response, and mean survival at one year was 74% (Mauro and Druker). In blast crisis patients, typically the most difficult group to treat, 64% of the 260 patients achieved either a hematologic or cytogenetic response while the median survival of 6.8 months compared favourably with the 3 month median survival for bone marrow transplant recipients (Mauro and Druker). As a result of its clear clinical benefits for the treatment of CML, in May, after a mere three months of review, the FDA approved STI571 for general use under the trade name of Gleevec. The speed of this decision reflects the amazing significance of Gleevec's application for the treatment of an often fatal disease.

Bibliography

Altmann, J.: *Military uses of Nanotechnology: Perspectives and Concerns*, Cambridge MA: Harvard University Press, 2005.

Attebery, B.: *Dust, Lust, and Other Messages from the Quantum Wonderland*, New York: Mercury Press, 2002.

Baird, D.: *Probing the History of Scanning Tunneling Microscopy*, New York: Mercury Press, 1999.

Berube, D.M.: *The Rhetoric of Nanotechnology*, New York: Mercury Press, 2001.

Braun, T.: *Nanoscience and Nanotechnology on the Balance*, Cambridge MA: Harvard University Press, 2008.

Bueno, O.: *Self-Reproduction and the Constitution of Nanophenomena*, New York: Mercury Press, 2001.

Carroll, J.S.: *Social Science Research Methods for Assessing Societal Implications of Nanotechnology*, Cambridge: Cambridge University Press, 2008.

Crow, M.: *Nanotechnology and Societal Transformation*, New York: Adenine Press, 2006.

Marshall, K.: *Atomizing the Risk Technology*, Cambridge MA: Harvard University Press, 2008.

Mayer, S.: *From Genetic Modification to Nanotechnology: The Dangers*, Cambridge: Cambridge University Press, 2005.

Mehta, M.D.: *Nanoscience and Nanotechnology: Assessing the Nature of Innovation in these Fields*, Chicago: University of Chicago Press, 2000.

Meyer, P.: *Does Science Push Technology*, London: Methuen, 1999.

Milburn, C.: *Nano/Splatter: Disintegrating the Postbiological Body*, New York: Mercury Press, 1998.

Mnyusiwalla, A.: *Nanotechnology*, Amsterdam: John Benjamins, 2005.

Nordmann, J.: *Discovering the Nanoscale*, Cambridge MA: Harvard University Press, 2007.

Schummer, J.: *Nanotech Challenges*, Berkeley: University of California Press, 2003

Seeman, N.C.: *Biomolecular Stereodynamics*, New York: Adenine Press, 2002.

Seeman, N.R.: *Molecular Structure: Chemical Reactivity and Biological Activity*, Oxford: Oxford University Press, 2006.

Suchman, M.: *Envisioning Life on the Nano-Frontier*, New York: George Braziller, 2002

Sweeney, A.E.: *The Promises and Perils of Nanoscience and Nanotechnology: Exploring Emerging Social and Ethical Issues*, London: J. M. Dent and Sons, 2007.

Tenner, E.: *Nanotechnology and Unintended Consequences*, New York: Lindisfarne Books, 2001.

Wald, C.A.: *Working Boundaries on the NANO Exhibition*, New York: Random House, 2005.

Weckert, J.: *The Control of Scientific Research: The Case of Nanotechnology*, New York: Random House, 2004.

Weil, V.: *Ethical Issues in Nanotechnology*, New York: Solomon R. Guggenheim Foundation, 2007.

Wolfson, J.R.: *Social and Ethical Issues in Nanotechnology: Lessons from Biotechnology and Other High Technologies*, New York: George Braziller, 2001.

Index

M

N

O

P

R

S

T

V